Soil Health Analysis

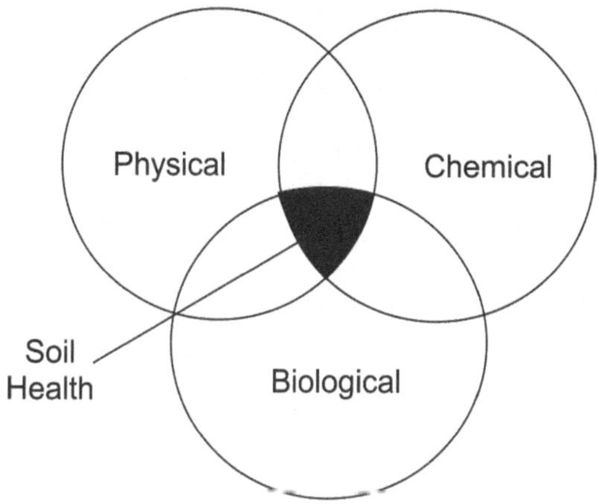

NIPA® GENX ELECTRONIC RESOURCES & SOLUTIONS P. LTD.
New Delhi-110 034

About Authors

Dr. Narendra Kumar Lenka is working as Principal Scientist in the field of Soil Science at the Indian Institute of Soil Science, Bhopal. He has about 19 years of experience in the field of Soil Science and Natural Resource Management. He is an alumni of the Indian Agricultural Research Institute (IARI), New Delhi. He has authored more than 60 research papers and his area of work includes soil physics and soil and water conservation, soil health assessment and soil quality indexing, soil resilience studies and climate change impact on agriculture.

Dr. Jyoti Kumar Thakur is working as Scientist in the field of Agricultural Microbiology at the Indian Institute of Soil Science, Bhopal. He is a post graduate and doctorate in Microbiology from the prestigious Indian Agricultural Research Institute (IARI), New Delhi. He has about 10 years of experience in the field of agriculture/soil microbiology and has authored about forty publications including research journals and book chapters. His specialization covers aspects related to soil microbial processes, biodiversity, PGPR and biofertilizers.

Dr. (Mrs.) Sangeeta Lenka is presently working as a Scientist in the field of Soil Science at the Indian Institute of Soil Science, Bhopal. She has more than 12 years of experience in the field of Soil Science and Natural Resource Management. She is an alumni of the Indian Agricultural Research Institute (IARI), New Delhi. She has authored about fifty publications including research journals, book chapters and technical articles. She has specialization in the fields of conservation tillage, soil carbon sequestration and climate change.

Soil Health Analysis

Narendra K. Lenka
Principal Scientist
ICAR-Indian Institute of Soil Science
Bhopal, Madhya Pradesh

Jyoti K. Thakur
Scientist
ICAR-Indian Institute of Soil Science
Bhopal, Madhya Pradesh

Sangeeta Lenka
Scientist
ICAR-Indian Institute of Soil Science
Bhopal, Madhya Pradesh

NEW INDIA PUBLISHING AGENCY
New Delhi – 110 034

NIPA® GENX ELECTRONIC
RESOURCES & SOLUTIONS P. LTD.

101,103, Vikas Surya Plaza, CU Block
L.S.C. Market, Pitam Pura, New Delhi-110 034
Ph : +91 11 27341616, 27341717, 27341718
E-mail: newindiapublishingagency@gmail.com
Web: www.nipabooks.com

For customer assistance, please contact
Phone: + 91-11-27 34 17 17
Fax: + 91-11- 27 34 16 16
E-Mail: feedbacks@nipabooks.com

ISBN: 978-81-19254-29-3

Composed and Designed by NIPA®.

Preface

Soil health deterioration is currently one of the key threats to sustainable agriculture and food security. The term soil health, which is often used interchangeably with the term soil quality, basically focuses on the critical functions that soils perform. As defined by Doran and Parkin, (1994), soil quality is "the capacity of soil to function within ecosystem boundaries to sustain biological productivity, maintain environmental quality, and promote plant and animal health". Soil health is a more comprehensive concept dealing with physical, chemical and biological properties of soil and its interrelations and takes a holistic view encompassing different functions of soil.

Despite novelty of the concept, routine soil analysis procedures mostly are limited to soil chemical and plant nutrient availability parameters, possibly due to the ease of analysis and the belief of a directly proportional relationship between plant nutrients and crop production. However, response of crop production to plant nutrients also depend upon soil health and physical structure. A healthy soil with good physical structure, friability and better water retention and transmission characteristics has a higher response to application of nutrients. Thus, to assess the health or quality of soil, additional set of parameters in addition to the parameters related to nutrient availability, are considered. In general, soil health parameters include physical, chemical and biological properties of soil, which reflect the overall health of soil.

Most of the Soil Analysis manuals currently available, primarily deal with basic chemical and physical parameters without giving an orientation to soil health assessment. However, certain indicator parameters such as soil carbon pools, soil aggregation behaviour, aggregate associated carbon content, biological activity etc indicate the health of a soil over a period of time. Soil microbes are vital for nutrient recycling and thus play a role in maintaining soil fertility. The recycling processes are brought about by several catabolic reactions mediated by enzymes secreted by microbes. Soil enzymatic activity gives an objective assessment of soil microbial health. Such indicator parameters though currently being used in research laboratories and by researchers, are not yet included in the framework of routine soil testing procedures.

This manual describes procedures for analysis of different soil health based indicator parameters including physical, chemical and biological parameters. The parameters which can be a soil health indicator, are only included in the manual. Further, for an assessment of soil health, it may not be essential to include all parameters given in the manual. Depending upon availability of facilities, a minimum data set of selected parameters can be considered. For keeping it simple, the manual is primarily sectioned into physical, chemical, carbon fractions and biological activity. The physical parameters such as bulk density, aggregation, water holding capacity etc which can be a part of soil health assessment framework, are detailed. In addition to basic chemical properties, procedures for analysis of macro and micronutrients are given. A detailed procedural description of different soil enzymes such as dehydrogenase, acid or alkali phosphatase, have been described.

The manual shall be useful for researchers, post graduate students, academicians and policy makers in the fields of Soil Science, Agronomy, Natural Resource Management and Forestry. The manual shall also be useful to the government bodies and planners in accomplishing objectives of soil health mission and soil conservation programs for maintaining optimum soil productivity.

Authors

Contents

Section 4: Soil Availability of Primary Plant Nutrients

Section 5: Soil Secondary and Micronutrients

6: Soil Biological Activity

Section 1: Common Soil Parameters

1

Collection and Preparation of Soil Sample

Soil sample collection is the first and most important step for soil quality analysis. A soil sample collected with due care and proper procedure is essential for getting reliable result of the analysis. The sample should represent the field and for practical purposes, an area of about 0.5 hectare is considered as one sampling unit. In case of conspicuous variation observed in the field in terms of texture, colour, slope etc, separate samples should be collected from areas differing in these attributes. In other words, sample should be collected from an area in the field with similar characteristics as seen through normal visual observation. Sampling should thus be avoided from areas like near trees, permanent vegetations, manure pits, manure piles or recently fertilized fields.

Soil Sampling procedure

1. For a particular field unit, one composite soil sample is prepared out of the samples collected from 7-8 spots distributed across the field unit. The entire field is covered by moving in a zig-zag manner from each sampling spot.

2. For each sampling point or spot, scrapping off the surface litter such as grasses or crop residues is done. Then, a "V" shaped pit of 15-20 cm length and of equal width is made. Depth of sampling depends upon the requirement and on the type of crop. For field crops like paddy, wheat etc, samples are collected from 15 cm depth. From the "V" shaped pits, take outslice of ½ inch thickness from both sides of the exposed pit surfaces from top to bottom.

3. The soil samples so collected from the 7-8 points are mixed thoroughly by hand and then is reduced to about 500 g by quartering process. The quartering process consists of a procedure in which the entire soil mass is divided into four quarters; the two opposite ones are discarded and the

remaining two are remixed. This process is repeated till about 500g soil is obtained.

Sampling tools

Soil sampling is generally done using an auger or spade or khurpi. Augers are basically of three types : screw, tube and post-hole auger. Tube auger is used for sampling a soft and moist soil, whereas, screw auger is preferred on a hard or dry soil. Post - hole augers are useful for sampling from wet areas like rice fields.

Sample processing

Soil samples collected are air dried under shade and then grinding is done using a wooden or porcelain made mortar and pestle. The sample is sieved through a 2.0 mm size sieve. The material that passed through 2.0 mm sieve is collected and stored in clean cloth or polythene bags for laboratory analysis. For certain analysis such as total organic carbon, 0.5 mm sieves are used.

Precautions in collection and storage of soil samples

- For micronutrients analysis, iron made sampling tools should be avoided. Rather, stainless steel auger should be used instead of iron made khurpi or auger.
- Soil samples should not be kept near to fertilizers, manures and pesticides.
- Containers previously used for storing fertilizers, pesticides, manures or chemicals should not be used for storing soil samples.

2

Determination of Soil Reaction (pH)

Soil pH is the measure of hydronium ion (H_3O^+) or more commonly hydrogen ion (H^+) activity in the soil suspension. Soil pH can be measured in soil-water suspension, saturation extract or in 0.01 M $CaCl_2$ solution. Saturation extract is used to identify specific soil problems such as salinity or sodicity. Measurement of pH using 0.01 M $CaCl_2$ solution gives more reproducible results than measured through soil-water suspension. It also provides a good approximation of field conditions as the electrolytic composition of 0.01 M $CaCl_2$ is considered similar to that of soil solution. However, among the three methods, soil-water suspension is most commonly used in research laboratories as well as for fertilizer recommendation as it is simpler and more convenient.

Measurement of pH in soil-water suspension

Soil pH can be measured in soil-water suspension, which may be 1.2 or 1:2.5 Soil-Water suspension.

Equipment and Reagents

1. Digital pH meter
2. Buffer solutions: pH 4.0, 7.0 and 9.2

Procedure

1. Take 20 g soil (2 mm sieved) in 100 ml beaker
2. Add 40 ml distilled water and stir the suspension with glass rod for five minutes.
3. Allow the suspension to stand for 30 minutes
4. Calibrate the pH meter using Buffer solutions of pH 4.0, 7.0 and 9.2.
5. Stir the soil water suspension with glass rod and record the pH.

Measurement of pH in 0.01 M $CaCl_2$ solution

Procedure

1. Take 20 g soil (2 mm sieved) in 100 ml beaker
2. Add 40 ml of 0.01 M $CaCl_2$ solution (1.470 g of AR grade $CaCl_2.2H_2O$ per litre)
3. Stir the soil-$CaCl_2$suspension with glass rod for 30 minutes.
4. Calibrate the pH meter using Buffer solutions of pH 4.0, 7.0 and 9.2.
5. Stir the suspension with glass rod just before taking the reading
6. Immerse the electrode and record the pH.

3

Determination of Soil Electrical Conductivity (EC)

(1:2 Soil-Water suspension)

Electrical conductivity (EC) is a useful indicator of soil salinity. Relationships have been developed between EC and salinity classes for a 1:1 soil: water suspension. With increase in salt content in a soil, the EC increases. Thus, for salt affected soils, measurement of EC is vital as accumulation of salts in the root zone is detrimental to plant growth. EC is a measure of the ability of a soil solution to conduct electrical current.

Equipment and Reagents

1) Digital Conductivity meter
2) 0.01 M potassium chloride (KCl) solution: Dissolve 0.7456 g dry KCl in distilled water and make up the volume to 1.0 litre. This solution has an electrical conductivity of 1.413 dS m^{-1} at 25°C.

Procedure

a. Take 20 g soil (2 mm sieved) in 100 ml beaker
b. Add 40 ml distilled water and stir the suspension with glass rod for 30 min.
c. Allow to settle the suspension until clear supernatant is obtained.
d. Calibrate the EC meter using 0.01 M KCl solution.
e. Record the EC of the clear supernatant.

Note: The clear extract after pH measurement can be used for EC measurement.

4

Cation Exchange Capacity (CEC)

Ion exchange is a reversible process by which a cation or anion held on the solid phase is exchanged with another cation or anion in the liquid phase. Soil mineral and organic colloidal particles have negative valence charges that hold dissociable cations. The power to retain cations at the surface of soil colloids is referred to as the cation exchange capacity (CEC). Cation exchange is a reversible reaction in soil solution, dependent upon negative charges of soil components arising from permanently charged or pH-dependent sites on organic matter and mineral colloid surfaces. The CEC is the sum of cations held by a kilogram of soil and is expressed in Cmol (p^+)/kg soil. In most agricultural soils, the cation-exchange capacity (CEC) is generally higher than the anion exchange capacity (AEC).

Reagent

1. **Neutral normal ammonium acetate solution** (1 N CH_3COONH_4 of pH 7): Dissolve 77.09 g of ammonium acetate in distilled water and make up the volume to one litre. Adjust the pH of the solution to 7 with ammonia solution or acetic acid.

2. **Sodium acetate solution** (1 N $CH_3COONa.3H_2O$ of pH 8.2): Dissolve 136 g of ammonium acetate in distilled water and make up the volume to one litre. Adjust the pH of the solution to 8.2 with NaOH solution or acetic acid.

3. **Ethyl alcohol 95%**

4. **Standard solution of Na (500 mg/L Na):** Dissolve 1.271 g of NaCI in distilled water and make up the volume to one litre.

5. **Working standard solution of Na (100 mg/L):** Dilute 200 ml of 500 mg/L stock solution to 1000 ml. Then prepare working standard solutions of Na (0, 20, 40, 60, 80 and 100 ppm) for standard curve.

Procedure

1. Weigh 4 g soil **(2 mm sieved)** and transfer it to 50 ml centrifuge tube.
2. Add 33 ml of normal sodium acetate solution.
3. Shake for 5 minutes on reciprocating shaker and then centrifuge at 3000 rpm until supernatant liquid is clear.
4. Decant the supernatant completely as possible and discard.
5. Repeat the above procedure (point 2 to 4) for 3 more times.
6. Now add 33 ml 95% ethanol and shake for 5 minutes on reciprocating shaker and then centrifuge at 3000 rpm until supernatant liquid is clear.
7. Decant the supernatant completely as possible and discard.
8. Repeat the above procedure (point 6 to 7) for 3 more times.
9. Now replace the adsorbed Na from the sample by extraction with three 33 ml portion of 1 N Ammonium acetate solution. Each time, shake for 5 minutes and centrifuge until supernatant is clear.
10. Decant the three supernatant liquid as completely as possible in to a 100 ml volumetric flask, bring to volume with 1 N Ammonium acetate solution and mix well.
11. Calibrate the flame photometer with the help of standard sodium solutions.
12. Take flame photometer reading of the sample aliquots and calculate the concentration of Na in mg/L.

Calculations

Weight of soil sample taken	=	4 g
Volume of the neutral normal CH_3COONH_4 solution added	=	100 ml
Dilution factor	=	100/4=25
Flame photometer reading of Na (mg/L)	=	R
(i) in mg/kg	=	R x 25=A
(ii) CEC (meq/100 g soil or Cmol(p^+)/kg soil	=	A/230

5

Exchangeable Sodium (Na^+) (Ammonium acetate method)

Exchanges sodium (Na^+) is an important parameter for salt affected soils. This is estimated by extraction with ammonium acetate and then measuring the extract in flame photometer.

Reagents

1. **Neutral normal ammonium acetate solution (1 N CH_3COONH_4 of pH 7):** Dissolve 77.09 g of ammonium acetate in distilled water and make up the volume to one litre. Adjust the pH of the solution to 7 with ammonia solution or acetic acid.

2. **Standard solution of Na (500 mg/L Na):** Dissolve 1.271 g of NaCl in distilled water and make up the volume to one litre.

3. **Working standard solution of Na (100 mg/L):** Dilute 200 ml of 500 mg/L stock solution to 1000 ml.

Procedure

1. Weigh 5 g soil **(2 mm sieved)** and transfer it to 150 ml conical flask.

2. Add 25 ml of neutral normal ammonium acetate solution.

3. Shake for 5 minutes on rotary shaker and filter through Whatman No.1 filter paper.

4. Prepare the standard solutions of sodium using working standard solution of Na (0, 20, 40, 60, 80 and 100 ppm).

5. Calibrate the flame photometer with the help of above standard solutions.

6. Take flame photometer reading of the sample aliquots and calculate the concentration of Na in mg/L.

Calculations

Weight of soil sample taken	= 5 g
Volume of the neutral normal CH_3COONH_4 solution added	= 25 ml
Dilution factor	= 25/5=5
Flame photometer reading of Na (mg/L)	= R
(i) in mg/kg	= R x 5 = A
(ii) Exchangeable Na^+(meq/100 g soil or Cmol/kg soil)	= A/230

Section 2: Soil Physical Parameters

6

Soil Texture (Bouyoucos Hydrometer Method)

Soil texture refers to the relative distribution of sand, silt and clay on a sample of <2 mm size. It is an intrinsic soil property and basically refers to the particle size distribution in a soil. A number of physical characteristics and chemical properties of a soil are influenced by texture or particle size distribution. To estimate the content of sand, silt and clay particles, the soil samples are first brought into a dispersed state by mechanical and chemical treatments so that different binding agents are removed. In the dispersed soil samples, the soil particles settle down at a differential settling rate according to their size. In laboratory, the common methods for analysis for particle size distribution are : Buyoucos hydrometer method and International pipette method. In this chapter, the procedure of Buyoucos hydrometer method is described.

Principle

This method is based on the principle of variable rate of settling of particles as per their size and as described by Stokes' Law. Large particles shall settle fast. The measurement of density of the suspension is done at different times using a hydrometer (ASTM 152H-Type hydrometer is preferred). The measurement in this hydrometer is based on a standard temperature of 20 °C and a particle density of 2.65 Mg m^{-3} and the units are expressed in grams of soil per liter.The principal source of error in this procedure is the incomplete dispersion of soil clays as the clay particles are cemented by agents such as organic matter, $CaCO_3$and iron oxides. Incomplete dispersion results in low values for clay and high values for silt and sand. Thus, soils with higher organic matter (>1.0%) should be first treated with 6% hydrogen peroxide (H_2O_2) for removal of organic carbon. The rate of sedimentation is also affected by temperature and the density of the dispersing solution.

Apparatus and Materials

- Glass cylinder of 1000-ml capacity
- Thermometer
- Hydrometer
- Mechanical stirrer with dispersing cup
- Stirring rod

Reagent

- **5% Sodium hexametaphosphate ($Na_6P_6O_{18}$)**

It is commonly known as Calgon solution and acts as the dispersing agent. For a 5% solution, dissolve 50 g of Sodium Hexametaphosphate in distilled water and make to 1000 ml.

- **Amyl alcohol**

Procedure

1. Take 100 g of air-dry soil sample **(2 mm sieved)** in a 500 ml beaker.
2. In case soil sample has more than 1.0% organic matter, treat the sample with 6% hydrogen peroxide (H_2O_2) to oxidize the organic matter. Skip this step if the sample has <1% organic matter.
3. Add 200 mL of distilled water and 100 mL of sodium hexametaphosphate solution.Stir the suspension well with the help of a glass rod and keep it for at least two hours followed by mechanical stirring for 10 minutes.
4. Transfer the contents to a measuring cylinder of 1000 mL capacity. In case of foam observed on top surface of the cylinder, add 2-3 drops of amyl alcohol.
5. Shake the contents of the measuring cylinder 4-5 times in an upside-down manner by putting palm of the hand or a rubber stopper tightly.
6. Keep the cylinder on the table for settling of the contents and note down the time.
7. Insert the hydrometer in the suspension and record the first reading at 40 seconds after keeping the cylinder on the table in stand-still position. Let us say, it is the reading – A. The hydrometer reading shows the amount of clay and silt suspended as by this time, sand particles settle down.

8. Take the second hydrometer reading (B) after 6 hours 52 minutes of initiation of settling or keeping the cylinder on the table. This reading shows the amount of clay suspended as by this time, sand and silt particles settle down.

9. Run a blank, without taking soil and use the same steps as mentioned above.

Calculation

Temperature and density correction

- Add 0.2 unit to the readings of the samples for every 1° F above 67° F, and subtract 0.2 unit for every 1° F below 67° F. Alternatively, for temperature of each degree above 20 °C, add 0.4 units and for temperature of each degree below 20 °C, subtract 0.4 units from hydrometer reading.
- Subtract the density of the blank at each reading, from the corresponding density readings for the samples.

$$\text{Per cent clay} = \frac{\textit{Corrected hydrometer reading at } 6 \textit{ hours } 52 \textit{ minutes} \times 100}{\textit{Weight of sample}}$$

$$\text{Per cent silt} = \frac{\textit{Corrected hydrometer reading at } 40 \text{ seconds} \times 100}{\textit{Weight of sample}} - \%\ \textit{clay}$$

Per cent sand = 100 – (% slit + % clay)

The results of sand, silt and clay are put on the textural triangle to find out the texture.

Reference

Bouyoucos, G.J. 1962. Hydrometer method improved for making particle size analysis of soils. Agron. J. 54:464-465.

7

Soil Bulk Density (Core method)

Soil bulk density (BD) or apparent density is determined in -situ by core method. Stainless steel or metallic cores of cylindrical shape with known height and diameter are inserted into the soil till the rim of the core. The soil core is then excavated and the soil sample is oven dried to get the dry weight. The ratio of the oven dry weight to the total volume of the core gives the bulk density of the soil.

Equipments/Apparatus required

1. Soil sampling core (made of stainless steel) with desirable height and diameter
2. Hot air oven
3. Khurpi or core auger

Procedure

1. Scrap the soil surface, clean it for grass or any residues and make the surface levelled
2. Keep the soil core on soil surface, put a wooden plank over it and slowly hammer it so that it inserts into the soil till the rim of the core
3. Excavate the core by digging the soil from four sides using a khurpi. Alternatively, core auger can be used if available
4. The soil sample collected may be transferred to a closed container or zip bag and taken to laboratory
5. Take the fresh weight of the sample and place it in a Hot air oven at 105 ^{o}C for minimum 24 hours or till attaining a constant weight
6. Record the weight of the dry sample

Calculation

BD (Mg m^{-3}) = M_s/V_t

Where, M_s = Mass of oven dry soil (g)

V_t = Volume of the cylindrical core (cm^3)=$\Pi r^2 h$ (r = radius of core; h = core height)

SI unit of BD is Mg m^{-3} (1.0 Mg m^{-3} = 1.0 g/cc)

Soil moisture per cent can be determined and expressed along with BD value

8

Soil Aggregate Analysis (Cambardella and Elliott ,1993)

Aggregation is a direct indicator of soil structure. A healthy soil is considered to have well aggregated structure. Though not routinely followed during soil testing and chemical analysis, aggregation is one of the major indicators of soil physical health. Wet sieve method (Yoder, 1936) is usually followed for soil aggregate analysis. This involves gently raising and lowering a nest of sieves in water to assess the stability of aggregates to rapid wetting. However, the Yoder's wet sieving method has been modified to make it simpler and currently, the method of Kemper and Rosenau, 1986 is being commonly used in research laboratories. One of the pre-requisites in these methods is the availability of wet sieving apparatus and further, these methods are laborious and time consuming. On the other hand, the method described by Cambardella and Elliott (1993) is considered very ideal because of its simplicity and is being widely used for aggregate bound soil carbon fraction study. The soil sample is wet sieved with a series of three sieves (2000, 250 and 53 micron size).

Aggregates are physically separated in four aggregate-size fractions:

i. Large macroaggregates (> 2000 µm in diameter)

ii. Small macroaggregates between 2000 -250µm in diameter

iii. Microaggregates between 250-53µm in diameter

iv. The mineral fraction <53 µm in diameter

Apparatus and reagents

- Sieve of 8 mm size
- Sieves of 2000, 250 and 53 µm (one each)
- Water tub
- Sodium hexa meta phosphate (0.5% or 5 g L^{-1})

Procedure

1. Collect moist soil samples from the field
2. Air dry and sieve it through 8 mm size sieves
3. Take 100 gram of air dry aggregates on the top of 2000 micron sieve and submerge it for 5 minutes in room temperature in deionized water.
4. Manually move the sieve up and down 3 cm with 50 repetitions during a period of 2 minutes.
5. After the two minutes period, the fraction that remained on the top of 2000 micron sieve are gently washed into an aluminum pan after discarding the floating organic matter.

Pour the water plus soil that passed through the 2000 micron sieve were poured on to the next size sieve (250 micron) and the step is repeated, but the floating material on the 250 and 53 micron sieves are retained.

The aggregates are oven dried at 50 °C and then weighed

Sand correction

For sand correction, the fractions in each size class are immersed in 0.5% sodium hexa metaphosphate for overnight and then pour it over 53 micron sieve. The material retained over 53 micron sieve is dried and is taken as the weight of sand.

Reference

Cambardella, C.A., Elliott, E.T., 1993. Carbon and nitrogen distribution in aggregates from cultivated and native grassland soils. Soil Science Society of America Journal, 57, 1071-1076.

9

Soil Aggregate Analysis (Dry sieving)

Aggregation is one of the basic physical characteristics of soil. Under natural conditions, primary soil particles, particularly clays, tend to join together to form secondary units called aggregates. Size distribution of soil aggregates is important because the size of the aggregates determines their susceptibility to water and wind erosion. Also, their size is important in determining the pore size distribution and regulating pore space in cultivated soils.

There are a variety of procedures being in use to test the stability of aggregates and aggregate size distribution. However, choice of the method for analysis of aggregate size distribution is determined by the purpose of analysis. The developed measures are relative and purpose specific. This chapter discusses dry aggregate analysis.

Dry Aggregate Analysis

Equipments

1. Rotary sieve shaker
2. Concentric sieves of various pore size to fit properly into the machine
3. Electronic balance

Sampling: For dry aggregate analysis, sampling is to be taken up when the soil is reasonably dry to avoid breakdown or change in structure. This is preferably done by a spade and gently keeping the collected soil in a tray.

Sieving: Soil sample (about 50 gm for a sieve of 20 cm diameter) is kept on the top sieve. A nest of sieves of various sizes (such as 8, 2, 1, 0.5, 0.25 mm) are kept on the rotary-sieve shaker with the sieve of biggest opening size placed on the top and that with the smallest at the bottom. The bottom sieve is fitted in a pan of same size and also the top sieve is covered with a pan of same size. The whole nest of sieves is fitted in the rotary-sieve shaker and bolted. The machine is then turned on for 5 to 10 minutes and the various sieves of dry aggregates remaining on different sieves are carefully collected and weighed.

Separation of stones

The estimates of dry aggregate analysis encounters mistake due to presence of small size stones similar to the size of aggregates. Hence, it is essential to separate small stone size materials from aggregates. This is done by grinding the dry aggregates in water and then passing it through the same sieve size. This way the stones of that size will be caught on that sieve itself. The remaining soil water suspension is then dried in a Hot Air Oven at 105 ^{0}C for 24 hours and weighed.

Note

The time of sieving should not be too short or unduly long, otherwise sieving of dry aggregates will not be complete or it will be over estimated due to breakdown. Therefore, proper timing for dry sieving should be carefully evaluated in each laboratory keeping in view the type of soils being handled in that laboratory.

Calculation

When the soil sample is being sieved for dry aggregates, a small amount of the same composite soil sample is kept in an oven for the determination of amount of moisture present in the soil. Thus, the whole soil sample taken for sieving can be converted on oven dry weight basis out of which the total amount of stones can be subtracted to get the actual amount of soil (oven dry basis) taken for dry sieving. Out of this oven dry sample, the percentage of different aggregate fractions (oven dried) can be calculated. If however, the amount of stone fraction in a soil is negligible, the dry aggregates caught on different size of sieves can be directly converted into percentage of the total amount taken.

10

Wet Sieving Procedure for Aggregate Analysis (Using Yoder's apparatus)

This procedure estimates aggregate stability to impact of rain drops and water erosion. Wet sieving procedure was originally proposed by Yoder, 1936. Later, the procedure was modified by Kemper and Rosenau, 1986.

Equipment

1. Yoder type sieve shaking electrically operated machine (3/4 inch stroke at 30 strokes per minute)
2. Concentric sieves of various sizes of pores to fit properly into the machine.
3. Weighing balance

Procedure

1. Soil samples should be preferably collected in cores. Aggregate size fractions are isolated by wet sieving using air-dry 8 mm sieved soil. Fill the Yoder's tank with water.
2. A set of sieves consisting six sieves of different diameters ranging from 4, 2, 1, 0.5, 0.25, 0.125 mm arranged in assembly at a descending order.
3. The dry screens are assembled into nests and then immersed in water in the Yoder's tank. The nests are then attached to the lift mechanism and minor height adjustments are made to bring the top screen of each nest into such a position that it just remains covered with water.
4. Air dried soil sample (100 gram) is kept on the top of first sieve (4 mm) and immersed in water at room temperature for 5 min.
5. The sieved are then oscillated vertically and rhythmically, so that water is made to flow up and down through the screens and the assemblage of aggregates.

6. The nests of sieves are mechanically raised and lowered through a distance of one inch at a rate of 30 oscillations per minute for a period of 10 minutes. At the end of 10 minutes period, the nest of sieves are removed carefully from the water, and aggregates retained on each sieve is back washed into an aluminum can and then dried in oven at 60 °C. After drying, weight of aggregate retained on each sieve is weighed.

Yoder's Apparatus

Sand correction

The results are corrected for the coarse primary particles retained on each sieve to avoid designating them falsely as aggregates. Therefore, sub sample of 5 g from each aggregate size is dispersed using sodium hexametaphosphate (5g/litre) and shaken for 16 hrs and passed through the lowest sieves (0.053 mm) for sand estimation. The mass retained on the respective size of sieves represents primary particles or sand fraction. Thereafter, sand fraction is deducted from the mass of aggregates to get realistic estimate of aggregates.

References

Kemper, W.D., Rosenau, R.C., 1986. Aggregate stability and size distribution. In: Klute, A. (Ed.), Methods of soil analysis. 2nd ed. Agronomy Monograph 9. ASA, Madison, WI, pp. 425–442.

van Bavel, C.H.M. (1949), "Mean weight diameter of soil aggregates as a statistical index of aggregation", Soil Science Society of America Journal, Vol.17, pp.416-418.

Yoder RE (1936) A direct method of aggregate analysis of soils and a study of the physical nature of soil erosion. J Am Soc Agron 28:337–351.

11

Mean Weight Diameter and Water Stable Aggregates

Wet sieving procedure separates soil into different size aggregates. However, for comparing the aggregate size distribution among management treatments and soil types, it is convenient to use indices. Mean weight diameter (MWD) and water stable aggregate percentage are some of the common indices used in aggregate stability studies.

Mean Weight Diameter (MWD)

This index was proposed by Van Bavel (1949) and is being used widely. The entire soil sample is passed through an 8 mm sieve prior to analysis and is subjected to wet sieving procedure as described before. The parameter which Van Bavel (1949) called the MWD is equal to the sum of the product of (a) the mean diameter (d_i) of each size fraction and (b) the proportion of the total sample weight (w_i) occurring in the corresponding size fraction. The summation is carried out over all 'n' size fraction including the one that passes through the finest sieve.

$$\text{MWD (mm)} = \sum_{i=1}^{n} d_i w_i$$

Illustration of MWD computation

The following example shows the steps in calculation and computation of MWD from a sample weight of 50 grams.

Observation and sample table for recording and computing soil aggregate size distribution and mean weight diameter

Sieve size class width (mm)	Mean diameter (d_i) (mm)	Weight of aggregate (g)	The proportion of the total sample weight	Sand (%)	Weight after sand correction (w_i)	d_iw_i
8.0 – 2.0	5	4	0.08	20	0.06	0.32
2.0-1.0	1.5	5	0.10	20	0.08	0.12
1.0-0.5	0.75	6	0.12	17	0.10	0.07
0.5-0.25	0.375	10	0.20	19	0.16	0.06
0.25-0.125	0.1875	11	0.22	15	0.19	0.04
<0.125	0.0625	14	0.28	5	0.27	0.02
					MWD (mm)	0.63

Water Stable Aggregates (WSA):

$$\text{WSA (\% of soil >0.25mm)} = \frac{\text{Weight of dry aggregates (>0.25 mm)} - \text{weight of sand x 100}}{\text{Weight of dry soil} - \text{weight of sand}}$$

Section 3: Soil Carbon and Fractions

12

Total Organic Carbon (TOC) (Dry combustion method)

In dry combustion technique, the combustion of soil sample (about 250 mg and **0.5 mm sieved**) is carried out at high temperature (about 1000 °C) through feeding in an inert boat in the electric furnace. The CO_2 produced from the oxidation of carbon present in the soil sample at high temperature is carried towards non dispersive infrared detector (NDIR) by a carrier gas where it is quantified and total carbon content is determined.

Equipments

TOC analyzer/CHN analyzer, weighing balance

Procedure

1. Weigh about 250 mg of soil sample **(0.5 mm sieved)** and transfer to sampling boat of the instrument (TOC analyzer/CHN analyzer)
2. Calibrate the instrument using sucrose after attaining the required conditions depending upon the type and make of the instrument (e.g. temperature, gas flow etc.).
3. Enter the exact weight of soil sample in the system and follow the instructions for feeding and removal of the sample boat.
4. Record the TOC content of the sample.

13

Determination of Soil Organic Carbon (Walkley & Black, 1934)

This is one of the commonly used method to estimate organic carbon content in soil. In this wet digestion method, soil organic matter is oxidised with a mixture of $K_2Cr_2O_7$ and concentrated H_2SO_4 utilizing the heat of dilution of H_2SO_4. The unused $K_2Cr_2O_7$ is back titrated with ferrous sulphate or ferrous ammonium sulphate.

Reagents

1. **Potassium dichromate (1 N $K_2Cr_2O_7$) solution**: Dissolve 49.04 g of analytical grade $K_2Cr_2O_7$ (dried at 105°C) in distilled water and make the volume to 1 litre.
2. **Concentrated sulphuric acid** (H_2SO_4)
3. **Ortho-phosphoric acid** (85%)
4. **Diphenylamine indicator:** Dissolve 0.5 g diphenylamine in a mixture of 100 ml concentrated sulphuric acid and 20 ml distilled water. Store it in an amber colour bottle.
5. **Ferrous ammonium sulphate(0.5 N $FeSO_4(NH_4)_2SO_4.6H_2O$) solution:** Dissolve 196.07 g analytical grade ferrous ammonium sulphate in distilled water, add 10 ml concentrated sulphuric acid and make volume to 1 litre with distilled water.

Procedure

1. Take 1.0 g of soil (0.5 mm sieved) in a 500 ml conical flask.
2. Add 10 ml of 1 N $K_2Cr_2O_7$ solution and shake gently to mix the contents.
3. Add 20 ml of concentrated sulphuric acid while swirling the flask slowly.

4. Keep the flask on a dry tile or asbestos sheet for 30 minutes and allow it to come to room temperature.
5. Add 10 ml of ortho-phosphoric acid, 200 ml of distilled water and 1-2 ml of diphenyl amine indicator solution. Shake vigorously to mix.
6. Then titrate against N/2 ferrous ammonium sulphate (FAS) solution till the colour changes from violet to bright green through blue.
7. Note the volume of ferrous ammonium sulphate solution used in titration.
8. Carry out a blank titration (without soil) in a similar manner.

Calculation

$$\text{Organic Carbon (\%)} = \frac{(B\text{-}S)\times(0.003)\times 100\times V}{B\times W}$$

In case of non-availability of TOC Analyser / CHN Analyser, total organic carbon (%) can be obtained by multiplying Walkley and Black organic carbon value with a factor of 100/76, as it in assumed that Walkley & Black method oxidises about 76% of soil organic, matter

Where

B- Volume of FAS used in blank titration (ml)

S- Volume of FAS used in sample titration (ml)

V- Volume of $K_2Cr_2O_7$ added (ml)

W- Weight of soil sample taken (g)

Reference

Walkley, A., and I. A. Black. 1934. An examination of the Degtjareff method for determining soil organic matter and a proposed modification of the chromic acid titration method. Soil Sci. 37:29–38.

14

Labile Carbon ($KMnO_4$-Oxidizable soil organic carbon) (Weil *et al.*, 2003)

Labile fractions of soil carbon (C) refer to the fraction of soil carbon that are easily altered by microbial activities. These fractions serve as sensitive indicators to changes in management induced soil quality and are grouped under active soil C pool. The method described by Weil *et al.* (2003) estimates changes in biologically active soil C using a dilute $KMnO_4$. This is an improvement over the method used by Blair *et al* (1995), where a very high concentration (0.333 M) of $KMnO_4$ was used.

Reagents

- **0.02 M $KMnO_4$:** Dissolve 3.16 g $KMnO_4$ in 900 ml distilled water and adjust the pH to 7.2 using 0.1 M NaOH and make the volume 1 lit. Store the solution in amber color bottle.
- **Sodium oxalate** ($Na_2C_2O_4$)
- **Sulphuric acid** (H_2SO_4)

Procedure

- Weigh 5 g **(0.5mm sieved)** soil sample in a centrifuge tube and add 20 ml 0.02 M $KMnO_4$.
- Shake for 2 minutes on a horizontal shaker (120 RPM) and then centrifuge it (4000-5000 RPM) to clear the supernatant.
- Take 2 ml of clear supernatant and make the volume to 50 ml with distilled water.
- Read the absorbance of the solution at 550 nm on spectrophotometer.
- To prepare standard curve, pipette 0, 1, 1.5, 2.0, 2.5 and 5.0 ml of the 0.02

M $KMnO_4$ and make the volume to 50 ml.

- Record the absorbance at 550 nm.
- Plot a curve for concentration Vs absorbance and record the slope of the curve.
- Also record absorbance of 0.02 M $KMnO_4$ solution and consider it as a reading for blank.

Standardization of $KMnO_4$

- Weigh 0.1 g $Na_2C_2O_4$ in a conical flask.
- Add 10 ml of 6N H_2SO_4.
- Heat the solution and titrate with $KMnO_4$ test solution.
- The end point is colourless to pink.

Calculation

$$\text{Exact molarity of } KMnO_4 = \frac{W \times 1000 \times 1000 \times 2}{M \times V \times 5}$$

Where,

W- Weight of $Na_2C_2O_4$ taken (g)

M- Molecular weight of $Na_2C_2O_4$ (g)

V- Volume of $KMnO_4$ used (ml)

Labile Carbon (mg kg^{-1}) = (B–S) × F × 9 × D

Where,

B- Absorbance of Blank

S- Absorbance of Sample

F- Slope of curve

D- Dilution factor

Reference

Weil, R.W., Islam, K.R., Stine, M., Gruver, J.B., Samson-Liebig, S.E. (2003) Estimating active carbon for soil quality assessment : A simplified method for laboratory and field use. American Journal of Alternative Agriculture, 18, 3-17.

15

Labile Carbon Fractions ($K_2Cr_2O_7$ Method) (Chan *et al.*, 2001)

Organic carbon present in soil can be very labile, labile or non-labile. This procedure measures the extent of lability of soil organic carbon through relative degree of oxidation by $K_2Cr_2O_7$.

Reagents

- **Potassium dichromate** (1 N $K_2Cr_2O_7$) solution:

  Dissolve 49.04 g of analytical grade $K_2Cr_2O_7$ (dried at 105°C) in distilled water and make the volume to 1 litre.

- **Concentrated sulphuric acid** (H_2SO_4)
- **Ortho-phosphoric acid** (85%)
- **Diphenylamine indicator:**

  Dissolve 0.5 g diphenylamine in a mixture of 100 ml concentrated sulphuric acid and 20 ml distilled water. Store it in an amber colour bottle.

- **Ferrous ammonium sulphate** (0.5 N $FeSO_4(NH_4)_2SO_4.6H_2O$) solution:

  Dissolve 196.07 g analytical grade ferrous ammonium sulphate in distilled water, add 10 ml concentrated sulphuric acid and make volume to 1 litre with distilled water.

Procedure

- Take 1.0 g of soil **(0.5 mm sieved)** in a 500 ml conical flask in three sets.
- For each set, add 10 ml of 1 N $K_2Cr_2O_7$ solution and shake gently to mix the contents.
- To one set, add 5 ml of concentrated sulphuric acid while swirling the flask slowly.

- To the second and third set, add 10 ml and 20 ml of concentrated sulphuric acid, respectively while swirling the flask slowly.
- Keep the flasks on a dry tile or asbestos sheet for 30 minutes and let them come to room temperature.
- Add 10 ml of ortho-phosphoric acid, 200 ml distilled water and 1-2 ml of diphenyl a mine indicator solution in all three sets. Shake vigorously to mix.
- Then titrate against N/2 ferrous ammonium sulphate (FAS) solution till the colour changes from violet to bright green through blue.
- Note the volume of ferrous ammonium sulphate solution used in titration for each set.
- Carry out a blank titration (without soil) in a similar manner.

Calculation

$$\text{Organic Carbon (\%)} = \frac{(\text{B-S}) \times (0.003) \times 100 \times \text{V}}{\text{B} \times \text{W}}$$

Where

B- Volume of FAS used in blank titration (ml)

S- Volume of FAS used in sample titration (ml)

V- Volume of $K_2Cr_2O_7$ added (ml)

W- Weight of soil sample taken (g)

Now,

Very labile pool (ppm) = (SOC content of first set) X 10000

Labile pool (ppm) = [(SOC content of 2nd set) – (SOC content of 1st set)] × 10000

Less labile pool (ppm) = [(SOC content of 3rd set) – (SOC content of 2nd set)] × 10000

Non labile pool (ppm) = [(TOC content (%) of soil as determined by dry combustion method) – (SOC content of 3rd set)] × 10000

Reference

Chan, K.Y., Bowman, A., Oates, A.L. (2001) Oxidizable organic carbon fractions and soil quality changes in an OxicPaleustalf under different pasture leys. Soil Science 166, 61-67

16

Particulate Organic Matter - Carbon (POM-C) (Cambardella and Elliott, 1992)

Particulate soil organic matter (POM) is one of the important soil organic matter (SOM) pools and matches the characteristics of a slow, decomposable or stabilized SOM pool. The POM fraction is considered as one of the indicators to changes in management and thus has been used in soil quality assessment studies.

Reagents

- **0.5% sodium hexametaphosphate** [$(NaPO_3)_6$]

  Dissolve 5 g of sodium hexameta phosphate in hot distilled water and make the volume to 1 litre

Procedure

- Weigh 20 g soil **(2 mm sieved)** in conical flask and add 100 ml 0.5% sodium hexametaphosphate.
- Shake for 15 hours on a reciprocal shaker at 120 reciprocations per minute.
- Pass the dispersed soil samples through 53µm sieve and collect the filtrate (silt+clay fraction).
- Dry the filtrate in an oven at 50 °C overnight and record the weight as (W_{S+C}).
- Determine the TOC content in Silt+Clay fraction.
- Also determine the TOC in bulk soil.
- Calculate the difference between the total bulk soil C and the C in the silt + clay size fraction by taking into account the weight of each.

- Find out the proportion of the POM-C as that of the total C.

Calculation

POM- C proportion (%) $= [(C_B \times W) - (C_{S+C} \times W_{S+C})]/(W \times C_B)$

Where,

W= Weight of soil sample (g);

W_{S+C}= Weight of silt+clay fraction (g);

C_B= Total organic carbon content in bulk soil (%);

C_{S+C}= Total organic carbon content in silt+clay fraction (%);

W_{POM} = Weight of POM Fraction (g);

C_{POM} = Particulate organic matter carbon in soil (%)

Reference

Cambardella, C.A. and Elliott,E.T., 1992. Particulate soil organic matter changes in a grassland cultivation sequence. Soil Science Society of America Journal, 56, 777-783.

17

Soil Respiration (Anderson, 1982)

Soil respiration is a good estimator of overall biological activity and thus is often used for describing soil quality. Soil respiration is estimated as the flux of CO_2 emitted from the soil surface. It represents the sum of CO_2 produced by root respiration and by heterotrophic decomposition. This procedure is a laboratory method where soil is incubated for a given period of days and the released CO_2 is trapped by sodium hydroxide (NaoH). This is also known as alkali trap method.

Reagents

- **1.0 M Sodium hydroxide**

  Dissolve 40g NaOH in distilled water and make the volume to 1 litre

- **1.0 M Hydrochloric acid**
- **Barium chloride solution (3 N)**
- **Phenolphthalein indicator**

  Dissolve 1 g of phenolphthalein in 100 ml 95% ethanol

Procedure

- Take 10-100 g soil sample**(2 mm sieved)** in a conical flask
- Maintain the moisture between 60-80% of field capacity
- Place a vial containing 10 ml 1.0 M NaOH inside the flask with the help of thread and seal the flask with rubber cork and paraffin wax
- Additionally, keep conical flasks without soil as blanks
- Take out the vials at periodic interval basis and add 1 ml 3N $BaCl_2$ solution.
- Titrate with 1.0 M HCl using phenolphthalein indicator till the colour changes from pink to white
- Record the volume of HCl consumed

Calculation

$$CO_2 - C \text{ mineralized (mg g}^{-1}\text{ soil)} = \frac{(B-S)\times N\times 6}{W}$$

Where,

B- Volume of HCl used in blank titration (ml)

S- Volume of HCl used in sample titration (ml)

N- Normality of HCl

W- Weight of soil sample taken (g)

Reference

Anderson JPE (1982) Soil respiration. In: Page AL, Miller RH, Keeney DR (eds) Methods of soil analysis, part 2. Am Soc Agron, Soil Sci Soc Am, Madison Wisconsin, pp 831-871

Section 4: Soil Availability of Primary Plant Nutrients

18

Soil Available Nitrogen (N) (Subbiah and Asija, 1956)

This is a distillation method proposed by Subbiah and Asija (1956), where the soil is distilled with alkaline $KMnO_4$ solution. The ammonia liberated is trapped in boric acid and then determined by titration against a standard acid. This procedure estimates the easily mineralizable nitrogen in soil.

Reagents

1. **Alkaline potassium permanganate (0.32%):** Weigh 3.2 g of potassium permanganate and dissolve in distilled water and make up the solution to one litre with distilled water.
2. **Sodium hydroxide (NaOH) (2.5 %):** Weigh 25 g of sodium hydroxide (NaOH) pellets in small quantity of distilled water after that make up to one litre by adding distilled water.
3. **Boric acid solution (2 %):** Dissolve 20 g of boric acid (H_3BO_3) in 800 ml of distilled water and 20 ml of mixed indicator solution. Make the volume up to one litre by adding distilled water. The pink colour appeares.
4. **Mixed indicator:** Dissolve 0.07 g of methyl red and 0.1 g of bromocresol green indicator in 100 ml of 95% pure ethanol and keep it in separate bottle.
5. **Liquid paraffin** (extra pure) or glass beads.
6. **Standard sulphuric acid (H_2SO_4) of N/100 :** Add 2.8 ml of concentrated sulphuric acid in small quantity of distilled water in a beaker. Then make up to one litre by adding distilled water. Take 100 ml of this solution and again make up to one litre. It gives N/100 (approximately) solution. This solution is titrated against primary standard solution (N/100) of sodium carbonate (Na_2CO_3) using methyl red as an indicator.

7. **Standard sodium carbonate (Na_2CO_3) of N/100:** Dissolve 0.53 g of sodium carbonate in distilled water and make up to one litre by addition of distilled water.

Method

1. Take 5 g soil sample **(2 mm sieved)** in a dry Kjeldahl (500 ml capacity) tube and add 10-15 ml distilled water. One milliliter liquid paraffin and some glass beads are also added in this flask. Then add 50 ml potassium permanganate (0.32%) in it.
2. Take 25 ml of boric acid solution (2%) in a conical flask. This flask is connected with another outlet of the condenser. This outlet tube of condenser must be dipped in the solution of conical flask.
3. Add 50 ml sodium hydroxide (NaOH) of 2.5% in Kjeldahl tube. Then the tube is immediately closed with rubber stopper. The burner is switched on immediately. About 100 ml of distillate is to be collected in the conical tube and colour of mixed indicator gets changed from pink to green. Remove the conical flask first and switch off the gas burner.
4. A blank sample should also run simultaneously without taking soil sample.
5. The solution of conical flask is titrated with standard sulphuric acid (N/100) solution till the pink colour changes to green and burette reading (volume of acid required for titration) is noted.

Calculation

Available Nitrogen (kg/ha) =

$[\{(X\text{-}Y) \times 14 \times \text{Normality of } H_2SO_4 \times 2 \times 10^6\} / \{\text{Wt. of soil} \times 1000\}]$

Wt. of the soil taken (g)	5 g
Volume (ml) of H_3BO_3 taken	25 ml
Volume (ml) of 0.01 N H_2SO_4 used for soil sample	*X ml*
Volume (ml) of 0.01 N H_2SO_4 used for blank	*Y ml*

19

Soil Available Phosphorus (P) (Olsen *et al.* 1954)

Estimation of plant available phosphorus content in soil has been attempted using several extractants. However, Bray's reagent and Olsen's reagent are the two common extractants used in India. Bray's reagent (0.03 N NH_4F and 0.025 N HCl) is used for acid soils, where, the fluoride ion complexes with Fe and Al in soil and thus releases the bound P. The Olsen's reagent (0.5 M $NaHCO_3$) is used for neutral-alkaline soils.

Reagents

1. **Standard phosphorus (P) solution**:
   i) **Standard P solution (100 mg/L Phosphorus):** Dissolve 0.4387 g of KH_2PO_4 in distilled water and make up the volume to one litre
   ii) **Working P solution (5 mg/L P):** Dilute 5 ml of 100 ppm P solution to 100 ml with distilled water
2. **N/2 $NaHCO_3$**: Dissolve 42 g of $NaHCO_3$ in distilled water and make up the volume to one litre after adjusting the pH to 8.5 with sodium hydroxide on a pH meter.
3. **Reagent A:** Dissolve 12 g Ammonium molybdate $[(NH_4)_6.Mo_7O_{24}.4H_2O]$ in 250 ml distilled water.Dissolve 0.2908 g ammonium antimony potassium tartarate $[K(SbO)C_4H_4O_6.1/2H_2O]$ in 100 ml water. Add these two solutions to 1000 ml of 2.5 M H_2SO_4, mix thoroughly and make to 2000 ml. Store in pyrex glass bottle in a dark and cool place.
4. **Reagent B:** Dissolve 1.056 g Ascorbic acid $(C_6H_8O_6)$ in 200 ml Reagent A and mix. This does not keep for more than 24 hours at room temperature. Prepare daily as required.

5. **2.5 M H_2SO_4:** Dilute 140 ml of concentrated H_2SO_4 to 1 litre.
6. **P-Nitrophenol indicator:** Dissolve 0.5 g p-Nitrophenol in 100 ml of distilled water.

Olsen's Method

a) Preparation of standard curve

1. To prepare the standard curve, pipette out 0, 0.5, 1, 2, 3, 4 ml of 5 mg/L P solution in six different 25 ml volumetric flaks. It gives 0, 0.1, 0.2, 0.4, 0.6 and 0.8 mg/L P.
2. Add 5 ml of N/2 $NaHCO_3$ solution and acidify with 2.5 N H_2SO_4 to it, mix and shake well until the evolution of CO_2 stops
3. Add about 10 ml of distilled water washing the neck of the flask.
4. Then add 4 ml of reagent B solution and make up the volume to the mark with distilled water and mix.
5. Blue colour of varying intensity will be developed in all flasks containing P solution.
6. Set the spectrophotometer at a wavelength of 730 nm and adjust the pointer of the galvanometer to read zero transmittance.
7. Then take the zero mg/L solution (blank) in the spectrometer tube, place it in the cuvette and set the pointer to read 100% transmittance or zero absorbance with 100 set knob.
8. Then read the colour intensity in each of the standard solution and record the absorbance.
9. Plot a curve showing relationship between concentration of P and the absorbance readings on a simple graph paper or Concentration against % T on semi-log graph paper.

b) Analysis of a test sample

1. Weigh 2g of soil sample**(2 mm sieved)** in a 100 ml conical flask.
2. Add a pinch of Darco-G 60 and 40 ml of N/2 $NaHCO_3$ solution.
3. Mix and shake the contents of the flask on an electric shaker at a constant speed for half an hour. Filter through Whatman No.1 filter paper.
4. In a similar way, prepare a blank with all reagents and no soil.
5. Take 5 ml of the filtrate in a 25 ml volumetric flask.

6. Add 2-3 drops of p-nitrophenol indicator. It develops yellow colour. Acidify the content by adding 5 N H_2SO_4 drop by drop until the content becomes colourless.
7. Then add 4 ml Reagent B, make the volume to 25 ml and shake the contents. Blue colour will develop.
8. Measure the intensity of blue colour after 30 min at 730 nm wavelength in spectrophotometer.
9. Similarly, pipette out 0, 0.5, 1, 2, 3, 4 ml of working P solution and follow the steps 6-8.
10. Determine the curve factor by plotting the concentration *vs* absorbance.
11. Calculate the P content in the sample.

Bray and Kurtz (1945) method

1. Weigh 2.0 g of 2 mm sieved air- dry soil in a 100 ml conical flask.
2. Add 20 ml of extractant for keeping soil to solution ration as 1:10.
3. Shake the suspension for 5 minutes on a reciprocating shaker.
4. Filter through Whatman No. 1 filter paper. If the filtrate is turbid, pour the filtrate back through the same filter paper.
5. Transfer 5 ml aliquot in a 25 ml volumetric flask. Add distilled water to 20 ml and then 4 ml of reagent B (described under Olsen's method). After 10 minutes, read the intensity of blue colour as described under Olsen' s method.

Calculations

Weight of the soil sample taken	2 g
Volume of N/2 $NaHCO_3$ solution (or the extractant) added	40 ml
First dilution	20 times
Volume of the filtrate taken for colour development	5 ml
Final volume made	25 ml
Second dilution	5 times
Total dilution	20 x 5 = 100 times
Absorbance of the test solution	A
Concentration of p read from standard curve against A	X mg/L
Available p in the soil (mg/L)	X x 100
Available p (kg/ha)	mg/L x 2.24

References

Bray RH, Kurtz LT. 1945. Determination of total, organic and available forms of phosphorus in soil. Soil Sci. 59:39-45.

Olsen SR et al. 1954. Estimation of available phosphorus in soils by extraction with sodium biocorbonate . Circular no 939, USDA, Washington DC

20

Soil Available Potassium (K) (Hanway and Heidel, 1952)

In soil system, potassium (K) exists as water soluble, exchangeable, non-exchangeable and lattice K. The plant available K refers to the sum of water soluble and exchangeable K. Estimation of available K is usually done using neutral normal ammonium acetate as the extractant.

Reagents

1. **Neutral normal ammonium acetate solution (1 N CH_3COONH_4 of pH 7):** Dissolve 77.09 g of ammonium acetate in distilled water and make up the volume to one litre. Adjust the pH of the solution to 7 with ammonia solution or acetic acid.
2. **Standard solution of K (1000 mg/L K):** Dissolve 1.91 g of KCI in distilled water and make up the volume to one litre.
3. **Working standard solution of K (100 mg/L):** Dilute 100 ml of 1000 mg/L stock solution to 1000 ml.

Procedure

1. Weigh 5 g soil **(2 mm sieved)**and transfer it to 150 ml conical flask.
2. Add 25 ml of neutral normal ammonium acetate solution.
3. Shake for 5 minutes on rotary shaker and filter through Whatman No.1 filter paper.
4. Prepare the standard solutions of potassium using working standard solution of K(0, 20, 40, 60, 80 and 100 ppm).

5. Calibrate the flame photometer with the help of above standard solutions.
6. Take flame photometer reading of the sample aliquots with necessary dilutions and calculate the concentration of K in mg/L.

Calculations

Weight of soil sample taken (WS)	= 5 g
Volume of the neutral normal CH_3COONH_4 solution added (VE)	= 25 ml
Dilution factor (X)	= Final volume/ aliquot taken
Reading of the flame photometer for the test sample	= FR
mg/L of K as read from the standard curve against	= Y
Available potassium in soil	
(i) in mg/L	= Y x (VE/WS) x X = A
(ii) in kg/ha	= A x 2.24

Section 5: Soil Secondary and Micronutrients

21

Available Sulphur (S) in Soil by Turbidimetric Method (Williams and Steinbergs, 1959)

In this method, plant available sulphur (S) is extracted with 0.15% $CaCl_2$ and the soluble sulphate is estimated on a colorimeter.

Reagents

1. **0.15% calcium chloride solution:** Dissolve 1.5 g of calcium chloride dihydrate ($CaCl_2.2H_2O$) in distilled water and dilute to one litre.
2. **Barium chloride ($BaCl_2$):** 30 to 60-mesh crystals: Grind barium chloride crystals until they pass through 30-mesh sieve and are retained on a 60-mesh sieve.
3. **Gum acacia solution (0.25%):** Dissolve 0.25 g gum acacia in distilled water and dilute to 100 ml.
4. **Standard sulphur solution (100 ppm S):** Dissolve 0.5434 g of the reagent grade potassium sulphate (K_2SO_4) in distilled water and dilute to one litre. Dilute this by 10 times to make a working standared of 10 ppm

Procedure

a) Preparation of standard curve

Pipette out 0, 2.5 25, 5.0, 10.0 and 15.0 ml of 10 ppm S solution in different 25-ml volumetric flasks to obtain 0,1,2,3,4 and 6ppm S, respectively.. Add in every flask 10 ml of 0.15% $CaCl_2$ solution and 1.0 g of 30-60 mesh $BaCl_2$ crystals. Swirl for one minute to dissolve the crystals and add 1 ml of 0.25% solution of gum acacia. Make up the volume in every flask with distilled water and shake well. Within 5 to 30 minutes after the development of turbidity read on a spectrophotometer at 340 nm using a blue filter. Plot a standard curve showing relationship between concentrations of S and turbidity/ absorbance readings.

b) Analysis of test sample

Weigh 10 g air-dried soil **(2 mm sieved)** and transfer it to a 150-ml conical flask. Add 50 ml of 0.15 % $CaCl_2$ solution and shake for 30 minutes on an electric shaker. Filter the suspension through Whatman No.42 filter paper. Pipette out 20 ml of the filtrate in 25-ml volumetric flask and proceed further as in case of standard curve. Run a blank with all the chemicals, except the soil. Find out sulphate concentration from the standard curve.

Calculations

Weight of soil taken	=	10 g
Volume of extractant added	=	50 ml
First dilution	=	5 times
Volume of aliquot taken	=	20 ml
Final volume	=	25 ml
Second dilution	=	1.25 times
Total dilution	=	5 x 1.25 = 6.25 times
Absorbance reading	=	A
S from standard curve against A (mg/L)	=	Y
Available Sulphur in soil (mg/L)	=	Y x 6.25 ppm
Available Sulphur content in soil (kg/ha)	=	ppm x 2.24

Reference

Williams CH, Steinbergs A. 1959. Soil sulphur fractions as chemical indices of available sulphur in some Australian Soils. Aust. J. Agric. Res. 10:340-352.

22

Available Boron (B) in Soil by Hot Water Soluble Method (Berger and Truog, 1939)

The hot water soluble method is the most common method to estimate available B content in soil

Reagents

1. **Azomethine-H:** Dissolve 0.45 g azomethine-H and 1.0 g L-ascorbic acid in about 100 ml de-ionized or redistilled water. If solution is not clear, it should be heated gently on water bath at about 30°C till it dissolves. A fresh solution should be prepared every week and kept in a refrigerator.
2. **Buffer solution:** Dissolve 250 g ammonium acetate in 500 ml de-ionized or redistilled water. Adjust the pH to about 5.5 by slowly adding approximately 100 ml glacial acetic acid, with constant stirring.
3. **EDTA reagent (0.025 M):** Dissolve 9.3 g EDTA in de-ionized or redistilled water and makes the volume up to 1 litre.
4. **Standard Boron solution:** Dissolve 0.571 g boric acid (H_3BO_3) AR grade in a small volume of deionized water and make volume to 1000 ml to obtain a stock solution of 100 mg B/L.

Method

(A) Preparation of standard curve

1. Take 5 ml of stock solution in a 100 ml volumetric flask and dilute to the mark. This solution contains 5 mg B/L
2. Take 0, 0.25, 0.50, 1.0, 2.0 and 4.0 ml of 5 mg B/L solution to a serried of 25 ml volumetric flasks.
3. Add 2 ml each of buffer reagent, EDTA solution and azomethine-H solution.

4. Mix the contents after each addition and allow tosand at room temperature for 30 minutes.
5. Make the volume to 25 ml with de-ionized or redistilled water and measure absorbance at 420 nm.
6. This will give reading for standard solution having B concentration 0, 0.05, 0.10, 0.20, 0.40 and 0.80 mg B/L.

(B) Extraction of Boron from soil

1. Weigh 25 g of soil **(2 mm sieved)**in a quartz flask or beaker (boron free glass).
2. Add about 50 ml of redistilled or de-ionized water and about 0.4 g of activated charcoal.
3. Boil the mixture for about 5 minutes and filtered immediately through Whatman filter paper No. 42.
4. Take 5 ml of the clear filtrate in a 25 ml volumetric flask and add 2 ml buffer reagent, 2 ml EDTA solution and 2 ml Azomethine-H solution.
5. Mix the contents thoroughly after the addition of each chemical.
6. Let the solution stand for 1 hr and allow developing colour and the volume is made to the mark.
7. Intensity of colour is measured at 420 nm.
8. The colour thus developed has been found to be stable up to 3-4 hours.

Calculations

Weight of the soil taken	= 25 g
Volume of extractant (water) added	= 50 ml
First dilution	= Two times
Volume of the filtrate taken	= 5 ml
Final volume of filtrate after colour development	= 25 ml
Second dilution	= 5 times
Total dilution	= 2 x 5 = 10 times
Absorbance of the soil solution as read from the spectrophotometer	= X
Concentration of B as read from the standard curve against X	= C mg /ml
Content of B in the soil (mg/L)	= C x 10 mg/ ml

References

Berger KC, Troug E. 1939. Boron determination in soil and plants. Ind, Eng. Chem. Anal. Ed. 11:540-545.

23

Soil Available Micronutrients (Fe, Mn, Zn and Cu) (Lindsay and Norvell, 1978)

This procedure uses Diethylene Tri-amine Penta Acetic acid (DTPA) as the chelating agent, which combines with free metal ions in soil solution and forms soluble complexes.

Reagents

1. **Extracting solution:** Dissolve 1.967 g Diethylene Tri-amine Penta Acetic acid (DTPA) and 1.470 g $CaCl_2.2H_2O$ (AR grade) in about 25 ml of de-ionized or double glass distilled water (DDW) by adding 13.3 ml TEA (Triethanolamine) followed by 100 ml more DDW. Transfer the solution to one litre volumetric flask giving 4-5 washings. Just before making up the volume, adjust pH to 7.3 with dilute HCl. This reagent has 0.005 M DTPA, 0.01 M $CaCl_2.2H_2O$ and 0.1 M TEA (Triethanolamine) adjusted to pH 7.3.
2. **Standard stock solutions** (1000 mg/L)

Analytical grade salts can be used to prepare stock standard solutions of different micronutrients. The quantity of the salts to be dissolved, its chemical formula, and concentration of respective stock solution is given below:

Element	Conc. of stock (mg/L)	Salt to be used	Quantity (g) to be dissolved in one litre of solution
Zn	1000	Zinc Sulphate ($ZnSO_4.7H_2O$)	4.398
	1000	Zinc oxide (ZnO)	1.244
Fe	1000	Ferrous sulphate ($FeSO_4.7H_2O$)	4.964
	1000	Ferrous ammonium sulphate($Fe (NH_4)_2 (SO_4)_2.6H_2O$)	7.028
Cu	1000	Copper sulphate ($CuSO_4.5H_2O$)	3.928
Mn	1000	Manganese sulphate ($MnSO_4.H_2O$)	3.076

The amount of salt as given above should be dissolved in a small volume of water followed by shaking after adding about 5 ml of 1:5 sulphuric acid. The content is then diluted to one litre with double distilled water.

Working standard solutions

Zinc (Zn): Transfer 10 ml of stock standard solution (1000 mg/L) to 100 ml volumetric flask and dilute up to the mark with DTPA to have a stock solution of 100 mg/L. Take 10 ml this solution (100 mg/L) to 100 ml volumetric flask and make up to the mark with DTPA to have a stock solution of 10 mg/L. Take 0, 1, 2, 4, 6 and 8 ml of standard stock of 10 mg/L to a series of 100 ml volumetric flasks and dilute each to the mark with DTPA. This will give standard solutions having Zinc concentration of 0, 0.1, 0.2, 0.4, 0.6 and 0.8 mg/L.

Iron (Fe), Copper (Cu) and Manganese (Mn) : Transfer 10 ml of stock standard solution (1000 mg/L) to 100 ml volumetric flask and dilute up to the mark with DTPA to have a stock solution of 100 mg/L. Transfer 0, 1, 2, 4, 6 and 8 ml of standard stock of 100 mg/L to a series of 100 ml volumetric flasks and dilute each to the mark with DTPA. This will give standard solutions having iron concentration of 0, 1, 2, 4, 6 and 8 mg/L.

Method

1. Weigh 20 g of soil **(2 mm sieved)** in 150 ml conical flask.
2. Add 40 ml of the DTPA extractant and shake for 2 hrs on mechanical shaker.
3. Filter the content through Whatman No. 42 filter paper, discarding first few drops.
4. Keep the filtrate to be analyzed for Zn, Cu, Fe and Mn with an atomic absorption spectrophotometer in polyethylene bottles.
5. Also keep a blank with each set following all the steps except the soil.
6. Take suitable quantity of aliquot in a 100 ml volumetric flask.
7. Make up the volume to 100 ml with glass distilled water.
8. Calibrate the AAS with the standard of the element to be analyzed.
9. Take reading of Zn/ Fe/ Mn/ Cu on AAS

Precautions

- Ensure that the de-ionized or double glass distilled water used is free of the micronutrient cations.

- The apparatus (glass/ polyethylene) to be used for the analysis must be thoroughly washed with acidified water and then with de-ionized water.
- Shaking time, DTPA concentration, pH and temperature during shaking influence the amount of Zn, Fe, Cu and Mn extracted by DTPA. The most suitable pH of extracting solution is 7.3, shaking time 2 hrs and temperature during shaking 25±1°C. So, do not forget to adjust the DTPA extracting solution pH to 7.3, shaking must be carried out at 25°C and the suspensions must be filtered immediately after the shaking time of 2 hrs.
- Before feeding the extracts, it should be ensured that they are not turbid otherwise, they may block the capillary of the AAS.

Calculations

Most of the modern AAS are calibrated to display the concentration of a micronutrient in mg/L directly in the soil extract. In such cases the concentration of the given micronutrient in the soil sample is calculated by multiplying the displayed reading by the dilution factor which is 2 in this case.

If the AAS displays the reading in absorbance then a standard curve has to be prepared for the known standards on a graph paper and the absorbance readings should be converted into concentration (mg/L) from the standard curve. The amount of the given micronutrient is then calculated as below

Weight of soil taken	20 g
Volume of DTPA extractant added	40 ml
Dilution	2 times
Absorbance shown by the AAS	A
Concentration of the micronutrient as read from the standard curve against A	C mg/L
Content of the micronutrient cation in the soil sample	C x 2 mg/kg

Reference

Lindsay WL and Norvell WA. 1978. Development of DTPA soil test for Zinc, Iron, Manganese and Copper. Soil. Sci. SOC. Am. J. 42:421-428.

Section 6: Soil Biological Activity

24

Techniques for Analysis of Soil Biological Parameters

Analytical methods for study of soil microbes

The analytical techniques for study of soil microbes are grouped under the following heads :

- Determination of form and arrangement of microorganisms in soil
- Isolation and characterization of soil microorganisms
- Detection of microbial activity in soil
- Determination of microbial biomass.

For the above, the different analytical methods used are described below.

1. Direct Count

Direct count is the method employed to estimate the counts of small arthropods, earthworms, nematodes, bacteria, fungi etc. Microscope with appropriate magnification and resolution are used to study organism of very small sizes. Direct counting by fluorescence microscopy can be a better option for direct count. For such microscopic study, the common stains used are fluorescein isothiocyanate (FITC), acridine orange (AO), ethidium bromide (EB) and europium chelate with a fluorescent brightener differential stain (DFS) for bacteria. Hyphae in agar films and on membrane filters can be stained by phenol aniline blue (PAB), whereas for metabolically active hyphae staining with fluorescent diacetate (FDA) gives better result. Stains for active bacterial cells include FDA or redox probes such as 2-(p-iodophenyl)-3-(p-nitrophenyl)-5-phenyl tetrazolium chloride (INT) or 5-cyano-2,3-ditoyl tetrazolium chloride (CTC). This procedure does not allow counting specific microbial species, and some of the stains used do not discriminate between living and dead microbial cells.

2. Culture dependent study

Traditionally microbial diversity has been quantified by counting techniques such as the plate count technique or the Most Probable Number (MPN) technique. The analysis of soil microbial communities using culture techniques employ variety of culture media designed to maximize the recovery of different microbial species. It has been estimated that less than 0.1% of the microorganisms found in typical agricultural soils are culturable using culture media formulations. Culture dependent study is inexpensive and can provide information on the active, heterotrophic component of the population. The limitations of culture dependent study include the difficulty in dislodging bacteria or spores from soil particles or biofilms, growth medium and growth condition selections. In addition, plate growth favours microorganisms with fast growth rates, numerically dominant species and those fungi that produce large numbers of spores.

3. Soil enzyme activity study

Since enzymes catalyze all biochemical transformations, measurements of soil enzyme activities are useful indicators of biological activity. Soil enzyme activity measurements have been used as indices of soil health as well as to understand how human activity is changing biogeochemical cycles in ecosystems. Some of the enzymes and associated nutrient transformation processes are listed in below.

Soil enzymes as indicators of nutrient cycling

Soil enzyme	Enzyme reaction	Indicator of microbial activity
Dehydrogenase	Electron transport system	C-cycling
β-glucosidase	Cellobiose hydrolysis	C-cycling
Cellulase	Cellulose hydrolysis	C-cycling
Phenol oxidase	Lignin hydrolysis	C-cycling
Urease	Urea hydrolysis	N-cycling
Amidase	N-mineralization	N-cycling
Phosphatase	Release of PO_4^-	P-cycling
Arylsulphatase	Release of SO_4^-	S-cycling

3.1 Fluorescein diacetate hydrolysis as a measure of soil biological activity

In soil more than 90% of the energy flow passes from the decomposition of organic matter by heterotrophic microbes. The decomposition of organic matter in soil is achieved by enzymes produced by several microbes inhabiting in the soil. These enzymes can also hydrolyse fluorescein diacetate. Hence estimation of fluorescein diacetate hydrolysis activity of soil serves as indicator of active microbial community in the soil. Fluorescein diacetate (3' 6'-diacetyl-fluorescein) is a fluorescein conjugated to two acetate radicals. This colourless compound is hydrolysed by both free (exoenzymes) and membrane bound enzymes, releasing a coloured end product, fluorescein. This end product absorbs strongly in the visible wavelength (490 nm) and can be measured by spectrophotometry.

3.2 Soil microbial biomass carbon (SMBC)

As soil microorganisms play an important role in the retention and release of nutrients and energy; any attempt to assess nutrient and energy flow in soil systems must take into account the role of soil microbial biomass.

A more easily applicable, non-subjective and replicable method is described for total microbial biomass determination in soil samples at a particular point of time. In the fumigation-extraction method, a direct measurement of C and other nutrients contained therein in microbial biomass is carried out. Overnight fumigation of chloroform is carried out to kill all the organisms in soil samples. The microbial biomass constituents released by $CHCl_3$ fumigation treatment can be extracted directly through chemical extractants. The readily oxidisable C contained in the extractant can be measured through standard chemical procedures.

The proportion of microbial biomass carbon in total soil organic carbon is known as microbial quotient. It gives insight into the capability of a soil to support microbial growth and soils with better quality is expected to maintain higher ratios of the microbial quotient. The metabolic quotient (qCO_2) is the ratio of basal respiration (BR) to the total microbial biomass carbon. The qCO2 indicates the efficiency by which soil microorganisms use C-resources in the soil. It is expected that stressed soils will provide higher qCO2 values than less-stressed soils.

3.3 Dehydrogenase assay

Lenhard (1956) introduced the concept of determining the metabolic activity of microorganisms in soil and other habitats by measuring dehydrogenase activity which was further refined by Casida et al, (1964) and is widely used for assessment of soil microbial activities. The technique involves the incubation of soil with 2,3,5-triphenyltetrazolium chloride (TTC) either in the presence or absence of added electron-donating substrates. Microbial dehydrogenase activity during this incubation results in reduction of the water soluble, colorless TTC to the water-insoluble, red 2,3,5-triphenyltetrazolium formazan. The latter is then extracted from the soil and read colorimetrically for quantification (Casida, 1977).

3.4 Phosphatase assay

A large pool of soil phosphorus exists in organic form and can be available to the plant only if converted to inorganic form. Phosphatases play a key role in phosphorous mineralization, and P cycling. They are ubiquitous in soil and have been studied extensively because they catalyze the hydrolysis of organic phosphomonoester to inorganic phosphorous, making it available for plant uptake. According to their optimum pH, phosphatases are classified as acid (orthophosphoric monoester phosphohydrolase, pH 6.5) or alkaline (orthophosphoric monoester phosphohydrolase, pH 11). Acid phosphomonoesterase activity is more important in acid soils than alkaline phosphomonoesterase activity, and vice versa in alkaline soils.

4. Physiological profiling based on carbon utilization pattern

This technique is based on utilization of different carbon sources by soil microbes and it determines the catabolic versatility of microbial community of an ecosystem. Community-level physiological profiles have been facilitated by the use of a commercial taxonomic system, known as the BIOLOG® system, which is currently available and has been used extensively for the analysis of soil microbial communities. Utilization of carbon substrate from 95 different carbon substrate provided is detected by the reduction of a redox dye, which results in a color change that can be quantified spectrophotometrically.

5. Phospholipid fatty acids (PLFA) analysis

Phospholipid fatty acids are potentially useful signature molecules due to their presence in all living cells and are found exclusively in cell membranes and not in other parts of the cell as storage products. Upon the cell death, membranes start degrading fast and the component phospholipid fatty acids are metabolized rapidly hence phospholipids can serve as important indicators of active microbial biomass as opposed to non-living microbial biomass. Some common signature phospholipid fatty acids of microbial community are given below.

Common signature phospholipids fatty acids of microbial community (Hill et al. 2000)

Common fatty acid signatures	
Common bacterial signatures	i15:0, a15:0, 15:0, 16:0, 16:1ω5, 16:1 ù 9, i17:0, a17:0, 17:0, 18:1 ω 7t, 18:1 ω 5, i19:0, a19:0
Aerobes	16:1 ω 7, 16:1ω7t, 18:1ω7t
Anaerobes	cy17:0, cy19:0
Sulfate-reducing bacteria	10Me16:0, i17:1ω7, 17:1ω6
Methane-oxidizing bacteria	16:1ω8c, 16:1ω8t, 16:1ω5c, 18:1ω8c, 18:1ω8t, 18:1ω6c
Cyanobacteria	18:2ω6
Protozoa	20:3ω6, 20:4ω6
Fungi	18:1ω9, 18:2ω6, 18:3ω6, 18:3ω3
Actinobacteria	10Me18:0
Microalgae	16:3ω3
Flavobacterium balustinum	i17:1ω7, Br 2OH-15:0

6. Genomic study

Research toward the taxonomic resolution and characterizing of the soil metagenomes is currently underway. The diversity of the microorganisms in soil can be estimated using one of two approaches, cellular cultures or molecular biology techniques.

Usually microbial diversity in soil has been analyzed by methods such as viable and cultivable methods. The molecular techniques generally involve extraction of nucleic acid, directly or indirectly, from soil. They are independent of culture and according to their sensitivity can detect species, genera, families or even higher taxonomic groups. Few of the commonly used methods are G:C ratio (Guanine : cytosine ratio), DNA : DNA hybridization, ribotyping, Fluorescent in Situ Hybridization (FISH), Ribosomal Intergenic Spacer Analysis (RISA), Multi-locus sequence typing (MLST), Random amplified polymorphic DNA (RAPD), Terminal Restriction Fragment Length Polymorphism (T-RFLP), Length Heterogeneity PCR and PCR-Denaturing Gradient Gel Electrophoresis (DGGE)/ Thermal gradient Gel Electrophoresis (TGGE).

7. Metagenomic approach

The collective genome of a total community in a given habitat is called metagenome. The term metagenomics was first used by Handelsman et al. (1998). Metagenomics enables us to study microorganisms by deciphering their genetic information from DNA that is extracted directly from communities of environmental microorganisms, thus bypassing the need for culturing, isolation and purification of culture. It offers insights into the evolutionary history as well as previously unrecognized physiological abilities of microbial communities specialized to live in a given environmental niche.

Approaches to Metagenomic Analysis (Handelsman, 2004)

- Isolation and purification of DNA from soil samples, often without prior cultivation of the organism present.
- Selection of suitable vectors and host system for cloning of the DNA obtained.
- Analysis of the cloned fragments using either sequence-based or functional screening approaches

Summary of molecular methods for soil microbial diversity studies and their applications. (Lynch et al, 2004)

Method	Type of information and resolution	Application in soil microbial analysis
Mole % G+C composition	Genetic community profile, overall community composition. Low resolution	Comparative analysis of overall changes in community composition
DGGE /TGGE sequencing of individualbands	Genetic fingerprinting of communities, affiliation of Predominant community members. Intermediate resolution	Comparative analysis of community structure, spatial and temporal changes in community composition
T-RFLP	Community composition, relative abundance of numerically dominant community members. Intermediate resolution	Comparative analysis of distribution of microbial populations, monitoring changes in community composition
PCR—ARDRA (Amplified Ribosomal DNA Restriction Analysis)	Genetic fingerprinting of simple communities,	• Comparative analysis of microbial population dynamics. • Diversity within phylogenetic or functional groups of microorganisms
RISA	Genetic fingerprinting of populations or phylogenetic groups. Simultaneously analysis of different microbial groups. Discrimination at species or group level. High resolution	• Comparative analysis of microbial population dynamics. • Diversity within phylogenetic or functional groups of microorganisms
FISH	Detection and specific counting of metabolic active microorganisms. Intermediate resolution	Comparative analysis of community structure. Detectionand identification of active cells. Direct phylogeneticinformation on community members

References

Casida, J.R. L.E. (1977). Microbial Metabolic Activity in Soil as Measured by Dehydrogenase Determinations. Appl. Environ.Microbiol. 34(6): 630-636.

Casida, L. E., Jr., D. A. Klein, and T. Santoro. 1964. Soil dehydrogenase activity. Soil Sci. 98: 371-376.

Handelsman, J. (2004). Metagenomics: Application of genomics to uncultured microorganisms. *Microbiology and Molecular Biology Reviews*. 68 (4):669–685

Lenhard, G. 1956. The dehydrogenase activity in soil as a measure of the activity of soil microorganisms. Z. Pflanzenernaehr. Dueng. Bodenkd. 73:1-11.

Lynch, J.M., Benedetti, A., Insam, H., Nuti, M.P., Smalla, K., Torsvik, V. and Nannipieri, P. (2004). Microbial diversity in soil: ecological theories, the contribution of molecular techniques and the impact of transgenic plants and transgenic microorganisms. Biol Fertil Soils. 40:363–385.

25

Enumeration of Soil Microbes by Dilution Plating Technique

Enumeration of microbial flora present in soil is done by dilution plating technique. The technique is based on the fact of diluting the source inoculums to the extent that after plating the known volume of inoculums on appropriate media, the colonies will appear in countable range and each colony will be assumed to have been derived from a single cell/colony forming unit. Three numerically dominant groups of soil microbes are bacteria, actinobacteria and fungi. The most commonly used culture media for their enumeration include

- Nutrient Agar-for heterotrophic bacteria,
- Actinomycetes isolation Agar or Kenknight and Munaier's agar for actinobacteria
- Martin's Rose Bengal Agar for fungi.

Materials required

Culture media, Autoclave, Laminar air flow, BOD Incubator, Petri plates, pipettes

Procedure

Serial dilution is prepared by adding 10g representative fresh soil sample to 90 mL sterile water blank (tap water is sterilized in autoclave) or 0.85% sodium chloride solution and labeled as 10^{-1}. From this dilution, 10 ml is subsequently transferred to next 90 mL water blank to make it 10^{-2}. Similarly, dilutions are prepared up to 10^{-6} for plating. From 10^{-4}, 10^{-5} and 10^{-6} dilutions, 100 µl of the inoculums are spread-plated on Petri plate containing pre-solidified nutrient agar media. For actinobacteria and fungi, from 10^{-2}, 10^{-3} and 10^{-4} dilutions, 100 µl of the inoculums are spread-plated on Kenknight & Munaier's agar and Martin's Rose Bengal Agar media. Three replications are maintained for each dilution along with an un-inoculated control of each media to detect any cross contamination. The plates are incubated at 28±2 °C for 2 days (for bacteria) 30±2°C for 5 days (for actinobacteria) and 25±2°C for 3 days (for fungi) in inverted position. The dilution where number of colonies are between 30-300 are considered for enumeration of viable cell count. Colonies are counted and expressed in terms of number of colony forming units/g of soil as

No. of cfu/g = No. of colonies / (dilution factor × volume of inoculum in mL taken for plating).

Observation

Bacterial colonies appear in varied shapes and sizes on plates, fungi appears like cottony growth on plate or fine mycelium can be seen radiating from colonies. Actinobacterial colonies are pigmented or bright white tough leathery with substrate or aerial mycelium.

Media composition

Nutrient Agar (pH 7.0)

Peptone	: 5.0 g
Beef extract	: 3.0 g
NaCl	: 5.0g
Agar	: 18.0 g
Distilled water	: 1000 mL

Kenknight and Munaier's medium(for Actinomycetes)

Dextrose 1.0g

Potassium di-hydrogen Phosphate 0.1g

Sodium nitrate 0.1g

Potassium chloride 0.1g

Magnesium sulphate heptahydrate 0.1g

Agar-Agar 18.0g

pH 7.2-7.5

Distilled water 1000 mL

Martin's Rose-Bengal medium (For fungi):

Dextrose 10.0g

Peptone 5.0g

Potassium di-hydrogen Phosphate 1.0g

Magnesium sulphate heptahydrate 0.5g

Rose Bengal (1% solution) 3.3ml

Agar-Agar 18.0g

Streptomycin Sulphate 50 mg*

Distilled water 1000mL

*Stock solution of Streptomycin Sulphate of 50mg/mL strength is to be prepared and filter sterilized. One ml of stock for 1.0 liter media is to be added after autoclaving and just before plating.

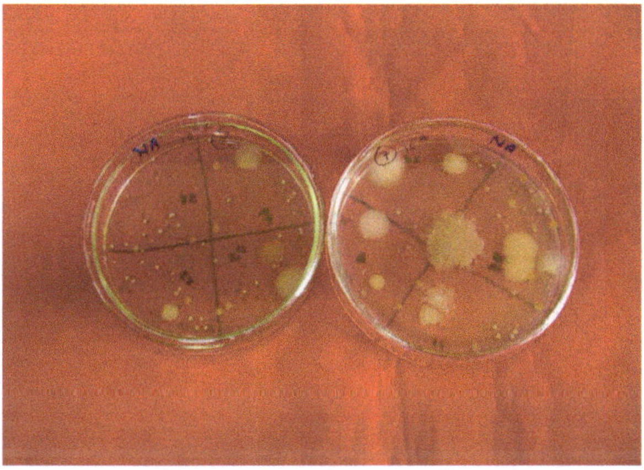

Bacterial colonies on Nutrient agar media

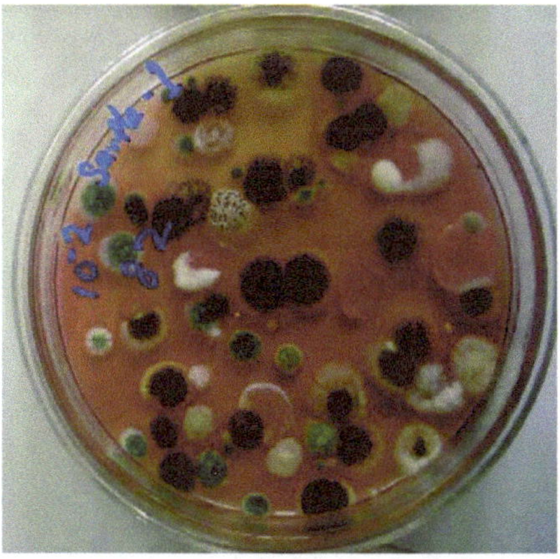

Fungal colonies on Martin's Rose-Bengal medium

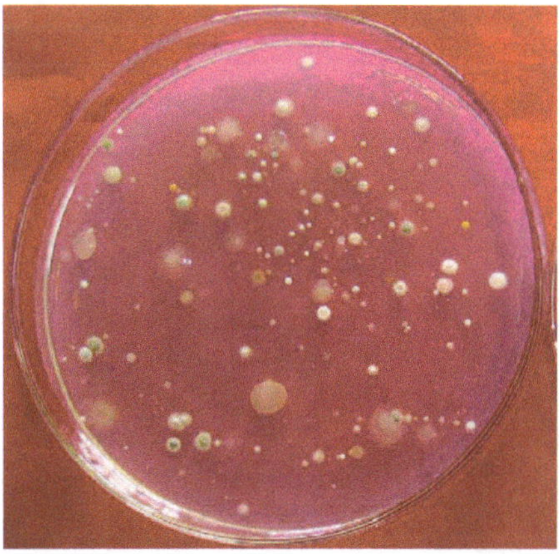

Actinobacterial colonies on Kenknight and Munaier's agar plate

26

Dehydrogenase Activity

Biological activity or index of a soil is the function of number of organisms present in soil and their physiological efficiency. The rate of respiration can be used as an index of the biological activity of soil as it reflects the physiological efficiency of the organisms. Monitoring of dehydrogenase, which are respiratory enzymes and integral part of all soil organisms, will give a measure of biological activity of soil at a given time.

During respiration biological oxidation of reduced compounds occur which is catalyzed by dehydrogenase. In the process TTC gets reduced to a pink coloured compound Triphenyl formazan (TPF) which can be quantitatively extracted by methanol and measured colorimeterically. The soil dehydrogenase activity in the current experiment was measured as per the methods of Cassida, (1977).

Reagents

- **2, 3.5 – triphenyl tetrazolium chloride** (TTC) solution (3%): Dissolve 3 g of TTC in 100 mL water. Store in an amber coloured bottle.
- **Methanol** (AR grade)
- **Glucose solution** (1%): Dissolve 1 g of glucose in 100 mL water.
- **Standard Triphenyl formazan** (TPF, 100 ppm): Dissolve 100 mg of TPF in 1000 mL methanol.

Procedure

1. Place representative 2 g air-dried soil in an air-tight amber coloured screw capped test tube (30 mL capacity) and add a pinch of calcium carbonate.
2. To each of the tube 0.5 mL of 3% TTC solution and 1ml sterile distilled water to saturate the soil. (Optional- Further 0.5 mL of 1% glucose solution can be added in each tube.) The bottom of the tube is tapped gently to drive out all trapped oxygen and a water seal is formed above the soil. A

blank is set up without TTC, instead 1.5ml distilled water is added to the soil.

4. Tubes are incubated at 37 ^{0}C for 24 h.
5. After incubation 10 mL of methanol is added and vortexed on vortex mixture. The tubes were left undisturbed for 10 minutes to settle the soil. Filter the supernatant and repeat the extraction with methanol. Make the volume to 25 ml with methanol in volumetric flask.
6. Read the Intensity of pink colour developed at 485 nm with a spectrophotometer.
7. Calculate the concentration of Triphenyl formazan formed from standard curve drawn in the range of 10 µg to 90 µg TPF mL^{-1} and express the result in terms of µg TPF formed per g soil per 24 hrs.

Note: Always dilute the a liquot if the OD of sample is beyond upper range of standard curve.

Reference

Casida, L. E., Jr., D. A. Klein, and T. Santoro. 1964. Soil dehydrogenase activity. Soil Sci. 98:371-376.

27

Fluorescein Diacetate (FDA) Hydrolysis

Fluorescein diacetate (3' 6'-diacetyl-fluorescein) is a fluorescein conjugated to two acetate radicals. This colourless compound is hydrolysed by both free (exoenzymes) and membrane bound enzymes, releasing a coloured end product, fluorescein. This end product absorbs strongly in the visible wavelength (490 nm) and can be measured by spectrophotometry. The enzymes responsible for FDA hydrolysis are plentiful in the soil environment. Non-specific esterases, proteases and lipases, have been shown to hydrolyse FDA. The hydrolysis of FDA by soil enzymes can be estimated by methods of Adam and Duncan (2001).

Reagents

- **60 mM potassium phosphate buffer pH 7.6.** Dissolve 8.7g K_2HPO_4 and 1.3 g KH_2PO_4 were in approximately 800 ml deionised water. The contents made up to 1 lit with deionised water. The buffer can be stored at 4°C and pH checked on day of use.
- **2:1 chloroform/methanol:** Chloroform (666 ml) (analytical grade) added to a 1lit volumetric flask. The flask is made up to 1L with methanol (analytical grade) and the contents mixed thoroughly.
- **1000 µg FDA/mL-stock solution:** Fluorescein diacetate (0.1 g) (3'6'-diacetyl-fluorescein) is dissolved in approximately 80 ml of acetone (analytical grade) and the contents of the flask is made up to 100 ml with acetone. The solution can be stored at -20°C.
- **2000 µg Fluorescein/ml stock solution:** Fluorescein sodium salt (0.2265 g) is dissolved in approximately 80 ml of 60 mM potassium phosphate buffer pH 7.6 and the contents made up to 100 ml with buffer.
- **20 µg Fluorescein/ml standard solution:** Stock solution (1 ml of 2000 µg fluorescein ml^{-1}) is added to a 100 ml volumetric flask and the contents

made up to the mark with 60 mM potassium phosphate buffer pH 7.6. Then 0-20 µg ml^{-1} standards are prepared from this standard solution by appropriate dilution in 60 mM potassium phosphate buffer pH 7.6.

Procedure

1. Place 2 grams of soil (fresh weight, sieved <2 mm) in a 50 ml conical flask and add 15ml of 60 mM potassium phosphate buffer pH 7.6.
2. Add 0.2 ml FDA from stock 1000 µg FDA ml^{-1} to start the reaction.
3. Prepare the blanks without the addition of the FDA substrate along with a suitable number of sample replicates.
4. Put stopper on the flasks contents are shaken by hand and placed in an orbital at 30°C for 20 minutes. To terminate the reaction, 15 ml of chloroform/methanol (2:1 v/v) is added in a fume hood. Stoppers are replaced on the flasks and the contents are shaken thoroughly by hand.
5. The contents of the conical flasks are then transferred to 50 ml centrifuge tubes and centrifuged at 2000 rev min^{-1} for approximately 3 min.
6. The supernatant from each sample is then filtered (Whatman, No 2) into 50 ml conical flasks and the colour developed is measured at 490 nm on a spectrophotometer.
7. The concentration of fluorescein released during the assay is calculated using the standard curve prepared with fluorescein sodium salt concentration ranging from 0-20µg/mL.

Reference

Adam, Gillian, and Harry Duncan. "Development of a sensitive and rapid method for the measurement of total microbial activity using fluorescein diacetate (FDA) in a range of soils." Soil Biology and Biochemistry 33.7-8 (2001): 943-951.

28

β-Glucosidase assay (Eivazi and Tabatabai, 1988)

Principle

β-Glucosidase (BG) plays a major role in the degradation of soil organic matter and plant residues. It catalyzes the hydrolysis of β-d-glucopyranosides, the final, rate-limiting step in the degradation of cellulose, the most abundant polysaccharide providing simple sugars for the soil microbial population. BG has been shown to be sensitive to changes in soil and residue management as well as an early indicator of changes in SOC before these changes are reflected in total or organic C analyses. The method is based on colorimetric determination of the p-nirtrophenol released by β-glucosidase when soil is incubated with buffered PNG solution and toluene.

Reagents

Modified universal buffer (MUB) stock solution: Dissolve 12.1 g of THAM, 11.6 g of maleic acid, 14.0 g of citric acid, and 6.3 g of boric acid (H_3BO_3) in 488 mL of 1 N sodium hydroxide (NaOH) and dilute the solution to 1 L with water. Store it in a refrigerator.

MUB, pH 6.0: Place 200 mL of MUB stock solution in a 500-mL beaker containing a magnet stirring bar, and place the beaker on a magnetic stirrer. Titrate the solution to pH 6.0 with 0.1 M hydrochloric acid (HCl), and adjust the volume to 1 L with water.

MUB, pH 11.0: Place 200 mL of MUB stock solution in a 500-mL beaker containing a magnet stirring bar, and place the beaker on a magnetic stirrer. Titrate the solution to pH 6.0 with 0.1 M NaOH, and adjust the volume to 1 L with water.

Calcium chloride ($CaCl_2$) (0.5 M): Dissolve 73.5 g of $CaCl_2.2H_2O$ in 700 ml of water, and dilute the volume to 1 L with water.

Standard p-nitrophenol solution: Dissolve 1.0 g of p-nitrophenol in 700 mL of water, and dilute the solution to 1 L with water. Store the solution in a refrigerator.

Tris (hydroxymethyl)aminomethane (THAM) buffer 0.1 M, pH 12 and~10: Dissolve 12.2 g of THAM in 700 mL of water, titrate the pH of thesolution to 10 by addition of approximately 0.1 M NaOH, and dilutewith water to 1 L.

p-Nitrophenyl β-D-glucoside (PNG) solution, 0.05 M: Dissolve 0.654g of PNG in about 40 mL of MUB pH 6.0, and adjust the volume to 50 mL with the same buffer. Store the solution in a refrigerator.

Procedure

- For each soil, three flasks are prepared (2 replications and 1 control).
- One g of soil (air-dried samples, <2 mm) is placed in each 50ml Erlenmeyer flask and 0.25ml toluene, 4 mL of MUB pH 6.0, and 1 ml of PNG solution is added to each flask. Swirl the flask to mix the contents. PNG solution is not added to the control.
- Flasks are incubated at 37°C for 1 h.
- After 1h, add 1 mL of 0.5 M $CaCl_2$, swirl and then add 4 ml of 0.1 THAM buffer pH 12, and swirl again.
- For controls: Follow same procedure but the addition of 0.5 mL of PNG solution after is the additions of 0.5 ml $CaCl_2$ and 2 ml of THAM buffer pH 12 (i.e., just before filtration of the soil suspension).
- Filter the soil suspension through a Whatman no. 2 folded filter paper into a pre-labeled test tube.
- Measure the intensity of the yellow color of the filtrate at 420 nm.
- If color intensity for a sample is high, dilute sample to within the limits of standard curve using THAM pH 10.

Preparation of Standard Curve

- Dilute 1 ml of standard p-nitrophenol solution to 100 ml in a volumetric flask using DI H_2O (deionised water).
- Pipette 0-, 1-, 2-, 3-, 4-, and 5-ml aliquots into a 50 ml Erlenmeyer flask and adjust the volume to 5 ml by addition of DI H_2O (this is equivalent to 0, 10, 20, 30, 40, and 50 μg of p -nitrophenol).

- Add 1 mL of 0.5 M $CaCl_2$ and 4 ml of THAM buffer pH 12.
- Filter suspension and measure the absorbance at 420nm.

Reference

Eivazi, F., and Tabatabai, M. A. (1988). Glucosidases and galactosidases in soils. Soil Biology and Biochemistry, 20(5), 601-606.

29

Phosphatase Activity in Soil

Acid and alkaline phosphatase enzymes are among the extensively studied phosphatases and classified so because of their optimum activities in acid and alkaline pH ranges, respectively. Acid phosphatase is predominant in acid soil while alkaline phosphatase is dominant in alkaline soils. The colorimetric assay of phosphatase is based on release of p-nitrophenol from hydrolysis of p-nitrophenyl phosphate by action of phosphatase enzyme in buffered soil solution (buffer pH 6.5 for acid and 11 for alkaline phosphatase assay).

Reagents

1. **Toluene :** Analytical grade
2. **Modified universal buffer (MUB) Stock solution:** Dissolve 12.1 g of THAM,11.6 g of maleic acid, 14.0 g of citric acid, and 6.3 g of boric acid (H_3BO_3) in 488 mLof 1 N sodium hydroxide (NaOH) and dilute the solution to 1 L with water. Store it in a refrigerator.
3. **MUB, pH 6.5:** Place 200 mL of MUB stock solution in a 500-mL beaker containing a magnet stirring bar, adjust to pH 6.5 with 0.1 M hydrochloric acid (HCl) while stirring, and make the volume to 1 L with distilled water.
4. **MUB, pH 11.0:** Place 200 mL of MUB stock solution in a 500-mL beaker containing a magnet stirring bar, adjust to pH 11.0 with 0.1 M NaOH, and make the volume to 1 L with water.
5. **p-Nitrophenyl phosphate solution (0.05M):** Dissolve 0.84g of p-Nitrophenyl phosphate tetrahydrate disodium salt in MUB (pH 6.5 for acid and 11 for alkaline phosphatase assay) and make the volume to 50 ml with same pH buffer.
6. **Calcium chloride (0.5 M):** Dissolve 73.5 g of $CaCl_2.2H_2O$ in 700 ml of water, and dilute the volume to 1 L with water.

Standard p-nitrophenol solution: Dissolve 1.0 g of p-nitrophenol in700 mL of water, and dilute the solution to 1 L with water. Store the solution in a refrigerator.

Procedure

- For each soil, three flasks are prepared (2 replications and 1 control).
- Place 1 g of soil (air-dried samples, <2 mm) in 50ml Erlenmeyer flask and add 0.25ml toluene, 4 mL of MUB (pH 6.5 for acid and 11 for alkaline phosphatase assay) and 1 mL of p-Nitrophenyl phosphate solution to each flask. Swirl the flask to mix the contents. p-Nitrophenyl phosphate solution is not added to the control.
- The flasks are incubated at 37°C for 1 h.
- After 1h, add 1 mL of 0.5 M $CaCl_2$, swirl and then add 4 ml of 0.5M NaOH, mix again.
- FOR CONTROLS: Follow same procedure but make the addition of 1 mL of p-Nitrophenyl phosphate solution after the additions of 0.5 ml $CaCl_2$ and 4 ml of NaOH (i.e., just before filtration of the soil suspension).
- Filter the soil suspension through a Whatman no. 2 filter paper into test tube.
- Measure the intensity of the yellow color of the filtrate at 420 nm.
- If color intensity for a sample is high, dilute sample to fall within the limits of standard curve using THAM pH 10.

Prepare standard curve as described for β-glucosidase assay and calculate the concentration of PNP released. Express the phosphatase activity in terms of μg PNP released/g soil/h.

References

Tabatabai, M.A. and Bremner, J. M. 1969. Use of p-nitrophenyl phosphate for assay of soil phosphatase activity. Soil Biology and Biochemistry, 1(4):301-307.

Evazi, Z. and Tabatabai, M. A. 1977. Phosphatase in soils. Soil Biology and Biochemistry, 9: 167-172.

30

Electrophoresis of DNA Molecule

Principle

Different ions migrate at different speed dictated by their size and by the amount of charges they carry. As a result, different ions can be separated from each other by electrophoresis. The magnitude of the electric force F_e exerted on the charged particle when subjected to electrophoresis is product of the charge q of the particle and the electric field E generated between the two electrodes.

$$F_e = q \times E \qquad (1)$$

From Stokes Law: For a spherical molecule of radius 'r', moving in viscous liquid with velocity 'v', the force acting on it is

$$F = 6\pi v r \eta \qquad (2)$$

$$q \times E = 6\pi v r \eta \text{ (from 1\&2)}$$

$$v = q \times E / 6\pi r \eta$$

Here, radius (size) is inversely proportional to the mobility. Hence, greater is the size, lower is the mobility in gel.

Electrophoresis in agarose or polyacrylamide gels is the most common method for separation of DNA molecules based on the size or length of nucleotide. Agarose gels can be used to separate molecules between 100 bp to 60 Kbp. Fragments larger than this size separated by a modification of electrophoresis termed pulsed field gel electrophoresis (PFGE). DNA gels are run on horizontal, gel electrophoresis as submerged gel where gel is totally immersed in buffer.

Once the system has been run, the DNA in the gel needs to be stained and visualised. The reagent most widely used is the fluorescent dye ethidium bromide. The gel is rinsed gently in a solution of ethidium bromide (0.5 mg cm^{-3}) and then viewed under ultraviolet light (300nm wavelength). Ethidium bromide (EtBr) is a cyclic planar molecule that intercalates the stacked base-pairs of DNA. The ethidium bromide concentration therefore builds up at the site of the DNA bands and under ultraviolet light the DNA bands fluoresce orange-red. Since DNA carriy net negative charge, it will move towards positive electrode.

Procedure

For running agarose gel, 0.8-0.9% agarose is prepared in 1X TAE buffer.

After cooling to 60 °C, 5 μL of 10 mg/ml EtBr for 100 ml gel (@0.05 μg/mL) is added and poured in tray to solidify it.

The solidified gel is placed in tank and TAE buffer 1x is poured into the tank high enough to cover the gel .

Add 1 μL loading dye per 5 μL sample (because the commercial dye is 6x) and electrophoresis is done at about 5v/cm voltage.

The band is visualized on UV trans-illuminator or gel documentation system.

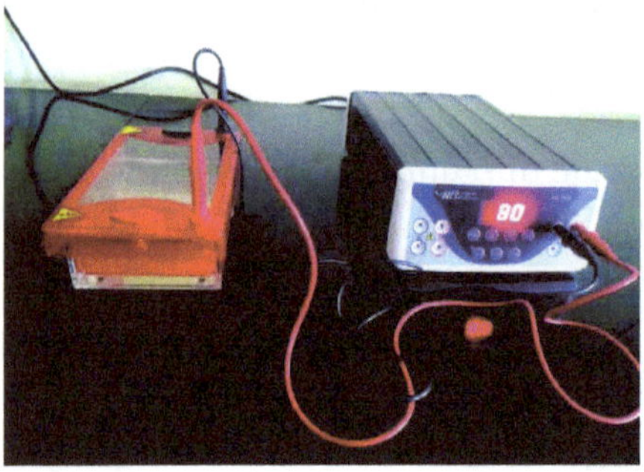

Plate: Horizontal Gel Electrophoresis system

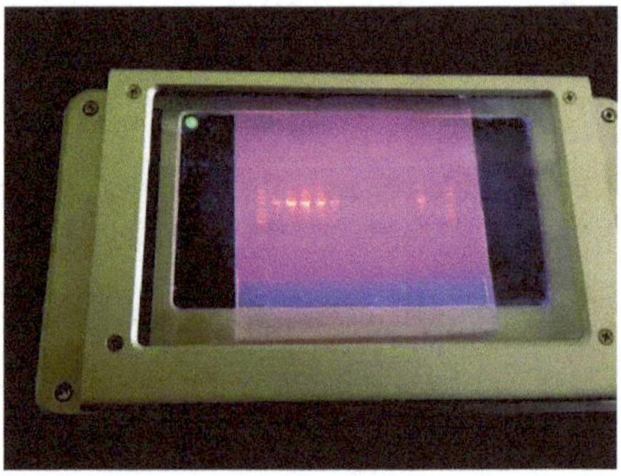

Plate. Gel stained in ethedium bromide (EtBr) to visualize DNA band in UV light

31

Important Terminologies used in Chemical Analysis

Mole

A **mole** is defined as the amount of chemical substance containing exactly $6.022\ 140\ 76 \times 10^{23}$ numbers of constitutive particles such as atoms, ions and molecules. This number is a fundamental constant named **Avogadro's number** in honor of Italian scientist Amedeo Avogadro. This constant is properly reported with an explicit unit of "per mole," and rounded version being 6.022×10^{23}/mol.

This definition has been adopted since November, 2018 revising the old definition based on the number of atoms in a sample of pure ^{12}C weighing exactly 12 g. Thus, as per the definition, 1 mole of any element contains the same number of atoms as 1 mole of any other element. However, the mass of 1 mole of a particular element is different from that of another element.

Molar mass

Molar mass of an element (or compound) is the mass in grams of 1 mole of that substance, a property expressed in units of grams per mole (g/mol).

Equivalent weight

It is the mass of one equivalent, which is the mass of a given substance that combines with or displaces a fixed quantity of another substance. By definition, equivalent weight is the mass of a substance which combines or displaces or contains 1.008 parts by mass of hydrogen, 8 parts by mass of oxygen or 35.5 parts by mass of chlorine. These values correspond to the atomic weight divided by the valence. If the equivalent weight is expressed in grams, then it is called gram equivalent mass. The equivalent weight of a compound can be calculated by dividing the molecular weight by the number of positive or negative electrical charges that result from the dissolution of the compound.

Methods of expressing concentration of a solution

Concentration of a solution is the amount of solute dissolved in a known amount of solvent. Each method of expression has certain advantages and use. Concentration of a solution is expressed in different ways, as mentioned below.

1. **Mass percentage** (% w/w) = $\frac{Mass\ of\ solute}{Mass\ of\ solution} \times 100$

A 5% (w/w) solution of NaCl means 5.0 g of NaCl in 100 g of solution

2. **Mass by volume percentage** (% w/v) = $\frac{Mass\ of\ solute}{Volume\ of\ solution} \times 100$

A 10% (%w/v) solution of NaCl means 10g of NaCl in 100 mL of solution.

3. **Volume percentage** (% v/v) = $\frac{Volume\ of\ solute}{Volume\ of\ solution} \times 100$

10% ethanol solution means 10 mL of ethanol in 100 mL of solution

4. Parts per million (ppm) and parts per billion (ppb)

In case the solute is present in trace quantity, it is expressed either as ppm or ppb. It can be on mass, volume or mass by volume basis as described in percentage above.

$$ppm = \frac{Number\ of\ parts\ of\ the\ component}{Total\ number\ of\ parts\ of\ the\ component\ in\ the\ solution} \times 10^6$$

$$ppb = \frac{Number\ of\ parts\ of\ the\ component}{Total\ number\ of\ parts\ of\ the\ component\ in\ the\ solution} \times 10^9$$

5. Molarity

Molarity is the number of moles of solute per liter of solution. In chemistry, the most commonly used unit for molarity is the number of moles per litre, having the unit symbol mol/L. A solution with a concentration of 1 mol/L is said to be 1 molar, commonly designated as 1 M. The SI unit for molar concentration is mol/m^3.

$$Molarity = \frac{Number\ of\ moles\ of\ solute}{Volume\ of\ solution\ in\ liter}$$

6. Molality

Molality is the number of moles of solute per kilogram of solvent. The SI unit for molar concentration is moles per kilogram. The major advantage of molality as a measure of concentration is that molality only depends on the masses of solute and solvent, which are unaffected by variations in temperature and pressure. On the other hand, molar concentration has a volume term, which is affected by pressure and temperature.

$$\text{Molality} = \frac{\text{Number of moles of solute}}{\text{Mass of solvent in kilogram}}$$

7. Normality

The normality of a solution is the concentration expressed as the number of equivalent weights or gram equivalents of the solute per liter of a solution. It is designated by the letter '*N*'. It is specifically used for acids and bases.

$$\text{Normality} = \frac{\text{Number of equivalent weights of solute}}{\text{Volume of solution in liter}}$$

8. Formality

It is the number of formula mass in grams present per litre of solution. In case, formula mass is equal to molecular mass, formality is equal to molarity. This method of expression is used for ionic compounds in which there is no existence of molecule. Mole of ionic compounds is called formole and molarity as formality.

$$\text{Formality} = \frac{\textit{Weight of solute } (gm)}{\textit{Formula weight of solute}} \times \frac{1}{\textit{Volume of solution } (L)}$$

9. Mole fraction

Mole fraction of any component in a solution is the ratio of number of moles of that component to the total number of moles of all components. For a solution with two components, A and B,

$$\text{Mole fraction of A} = \frac{nA}{nA + nB} \text{ and mole fraction of B} = \frac{nB}{nA + nB}$$

Where, nA and nBare the number of moles of component A and B, respectively

PREPARATION OF REAGENT SOLUTION

There are two basic methods for preparing reagent solutions – (1) preparing stock solution, and (2) preparing solutions by dilution. A stock solution is prepared by putting an exact mass of a pure solid or exact volume of a pure liquid and diluting it to a known volume.

For example, to prepare 500 mL of 0.10 M of NaOH, the amount of NaOH required is

$$= \frac{0.10\, mol\, NaOH}{L} \times \frac{40\, g\, NaOH}{mol\, NaOH} \times 0.50\ \text{L} = 2.0\ \text{g}$$

In the laboratory, solutions are often prepared by diluting concentrated stock solution. A known volume of stock solution is transferred to a new container and made up to a new volume. As the total amount of solute remains the same before and after dilution, the reagent preparation is done by using the formula

$$C_1V_1 = C_2V_2$$

Where,

C_1 is the concentration of stock solution

V_1 is the volume of stock solution required in L

C_2 is the concentration of diluted solution

V_2 is the volume of dilute solution required in L

For example, to prepare 500 mL of 0.10 M of NH_3, what is the amount of stock solution of concentration 14.8 M will be required. Substituting the known volumes in the equation, we get :

$14.8 \times V_1 = 0.5 \times 0.1$

Thus, $V_1 = 0.0033\ \text{L} = 3.3\ \text{mL}$

STANDARDIZATION OF REAGENT SOLUTION

Chemical reagents are manufactured and marketed in different grades of purity. In general, the purest reagents are marked as "analytical reagent " or AR-grade. Other labels are "LR", meaning laboratory, and "CP", meaning chemically pure. For microbiological analysis, reagents of higher purity with MB grade, meaning Molecular Biology grade are used. The strength of chemicals is usually expressed in terms of percentage, normality or molarity.

Buffer Solution

Precisely, it is known as pH buffer or hydrogen ion buffer. It is an aqueous solution consisting of a mixture of a weak acid and its conjugate base, or vice versa. For instance, a mixture of acetic acid (CH_3COOH) and sodium acetate (CH_3COONa) constitute a buffer. In this case, CH_3COOH is weakly ionized whereas, CH_3COONa is completely ionized.

Table : Strength of commonly used acids and alkalis

Reagent/chemical	Formula weight	Specific gravity	Normality	Molarity	Reagent % (w/w)	Volume (mL) required for 1 L of 1M solution
Acetic acid, glacial (CH_3COOH)	60.05	1.05	17.4	17.4	99.7	57.5
Ammonium Hydroxide (NH_4OH)	35.05	0.90	14.8	14.8	57.0	69.0
Hydrochloric acid (HCl)	36.46	1.19	12.1	12.1	37.0	82.5
Hydrofluoric acid (HF)	20.01	1.18	28.9	28.9	49.0	34.5
Nitric acid (HNO_3)	63.01	1.42	15.8	15.8	70.0	63.0
Sulphuric acid (H_2SO_4)	98.08	1.84	36.0	18.0	96.0	55.5
Perchloric acid 70% ($HClO_4$)	100.46	1.67	11.7	11.7	70.0	85.5
Phosphoric acid (H_3PO_4)	98.0	1.70	44.4	14.8	85.0	67.5

Some important terms that are often used in laboratory for chemical analysis are defined/explained below.

Buffer of a particular pH can be prepared if the acid dissociation constant of the acid and base dissociation constant of the base is known, so that the ratio of salt and acid or salt and base can be regulated. The pH of a buffer solution changes very little even upon addition of a strong acid or base. In biological analysis, buffer solutions are needed to keep the correct pH for enzymes. The major characteristics of a buffer solution are :

- The pH of a buffer does not change on dilution
- The pH of a buffer solution changes slightly upon addition of a small amount of acid or base.
- The pH of a buffer does not change upon standing for long

Titrations

Titration is a common laboratory method of quantitative chemical analysis that is used to determine the concentration of a reagent. This is also known as volumetric analysis since volume measurements are the integral part of such analysis. The reagent, concentration of which is known, is called titrant or titrator, while the reagent of unknown concentration is called titrand or titrate. The volume of titrant reacted is known as titration volume. The completion of the reaction is assessed with the help of an appropriate indicator.

There are different types of titrations, such as,

- Acid-base titration
- Redox titration
- Gas phase titration
- Complexometric titration
- Zeta potential titration

Among the above types, acid-base and redox titrations are most common.

1. Acid-base titration

An acid-base titration is a quantitative analysis procedure where an acid or base of known concentration neutralizes an acid or base of unknown concentration and the end point of which can be monitored through change in color of the solution upon added with an acid-base indicator.

Thus, before initiating a titration, one must choose a suitable indicator. As an indicative guide, the following can be used for selecting an indicator.

Indicator	Range of color change (pH)	Color on acidic side	Color on basic side
Methyl violet	0.0 -1.6	Yellow	Violet
Bromophenol blue	3.0 - 4.6	Yellow	Blue
Methyl orange	3.1 - 4.4	Red	Yellow
Methyl red	4.4 - 6.3	Red	Yellow
Litmus	5.0-8.0	Red	Blue
Phenolphthalein	8.3 -10.0	Colorless	Pink

2. Redox titration

This titration is based on reduction-oxidation reaction between an oxidizing agent and a reducing agent. To check the end point, a redox indicator is used. The most common example is the wet oxidation method for determination of soil organic carbon, where potassium dichromate ($K_2Cr_2O_7$) is the oxidizing agent and diphenyl amine or ferroin is used as the redox indicator.

Indicators

A substance that indicates the end point on completion of the reaction is called an indicator. The most commonly used indicators in volumetric analysis are:

- Internal indicators: They take part in the reaction and after end of the reaction, indicate the end point by change of color of the solution. Example : Methyl orange, phenolphthalein, methyl red
- External indicators: They never take part in the original chemical reaction, but after the completion of chemical reaction, indicate the end point by change of color of the solution. Example : Potassium ferricyanide
- Self – indicator (This is not truly an indicator but is one of the reacting species of the titration, either titrant or titrate. The best example is $KMnO_4$. After completion of titration, the color of the solution becomes colorless but with addition of extra $KMnO_{4,}$ the color changes to light pink.

Standard solution

A solution containing precisely known concentration of an element or substance is known as standard solution. Standard solutions are used to determine the concentration of other reagents or chemicals.

Primary standard

It is typically a reagent which can be weighed easily and is pure to the extent that, its weight represents the number of moles of substances it contains. A primary standard should have

- high purity
- low reactivity
- low hygroscopicity
- high equivalent weight so that error is minimum during mass measurements
- non-toxic.

Common examples of primary standards are sodium carbonate and Potassium hydrogen phthalate. Sodium carbonate is used for standardization of aqueous acids such as HCl, H_2SO_4 while potassium hydrogen phthalate is used for standardization of aqueous base.

Secondary standard

A secondary standard is a solution, concentration of which is measured by standardization or titration with a primary standard. The main reason for making a secondary standard solution is to prepare a solution of an exact concentration. For example, concentration of hydrochloric acid (HCl) vary concentration with time. Hence, a standard HCl solution is prepared by taking an approximate concentration of HCl and titrate it with a suitable primary standard solution, such as sodium carbonate. An exact concentration of HCl is then calculated after the titration is complete. Some secondary standards are sulphuric acid, hydrochloric acid, sodium hydroxide.

Note

Strength of standard solutions of all the reagents required in a laboratory must be checked periodically, or fresh reagents should be prepared before analysis.

Standardization of hydrochloric acid (HCl)

Let us say, we need to prepare a standard of 0.2 N HCl. The concentrated HCl has the normality of about 11.0 N. Thus, to prepare one litre of 0.2 N HCl, it is diluted about 50 times. Thus, the steps are :

- Take 20 ml of concentrated HCl and make it to 1.0 litre by dilution with distilled water.
- Titrate this solution against 0.2 N Na_2CO_3 (primary standard) using methyl orange as indicator.
- The colour changes from pink to yellow when the acid is neutralized.
- Suppose 10 ml of acid and 11 ml of Na_2CO_3 are consumed in the titration, then :

$10 \times N_1 = 11 \times 0.2$

$N_1 = 2.2/10 = 0.22$

Thus, the normality of the prepared HCl is 0.22

Standardization of sulphuric acid (H_2SO_4)

Suppose, we need to prepare one litre of 0.5 N H_2SO_4. The normality of concentrated H_2SO_4 is 36. Then, the steps shall be :

- Using the $V_1N_1 = V_2N_2$ concept, we get, $0.5\times 1000 = 36 \times V_2$

  Thus, $V_2 = 500/36 = 13.88$ ml = 15 ml (approximately)

- Take about 900 ml of distilled water, add 15 ml of concentrated H_2SO_4 to it and then make up the volume to one litre by adding more distilled water
- Titrate this solution against 0.5 N Na_2CO_3 (primary standard) using methyl orange as indicator.
- The color changes from pink to yellow when the acid is neutralized.
- Then, using the $V_1N_1 = V_2N_2$ formula, find out the exact normality of H_2SO_4.

Annexures

Annexure – I

Some conversion relationships

1.0 ppm	=	1.0 mg /L (solid per liquid)
	=	1.0 µg /mL (solid per liquid)
	=	1.0 µL /L (liquid per liquid)
	=	1.0 µg /g (solid per solid)
1.0 %	=	1.0 ppm × 10^4 or 10,000 ppm
	=	
Soil organic matter content (%)	=	Soil organic carbon (%) × 1.724 (100/58)
Protein (%)	=	N content (%) × 6.25
Salt concentration	(mg/lit) or ppm	Electrical conductivity (dS/m) × 640
	milli equivalent/lit	Electrical conductivity (dS/m) × 10
Osmotic pressure (atm)	=	Electrical conductivity (dS/m) × 0.36
P_2O_5	=	P × 2.29
P	=	P_2O_5 × 0.436
K_2O	=	K × 1.20
K	=	K_2O × 0.83

Annexure – II

List of some suppliers of chemicals

Sl.No.	Name of the manufacturer	Brand	Details
1	M/s Hi-Media Laboratories Pvt. Ltd., Mumbai	Hi-Media	23, Vadhani Industrial Estate, L.B.S. Marg, Mumbai - 400 086, India. Phone : 91-22-2500 1607, 6116 9797 Fax : 91-022-2500 5764, 2500 2468, 2500 2286 info@himedialabs.com
2	M/s Merck Life Science Pvt. Ltd., New Delhi	Merck	A-62, Naraina Industrial Area, Phase-1New Delhi - 110 028 Phone : +91 11 3350 0600 / 0660 Fax : +91 11 3350 0666 E-mail: nitika.rishi@external.merckgroup.com
3	M/s Thomas Baker Pvt. Ltd, Mumbai	Thomas Baker	4/86, Bharat Mahal, Marine Drive, Mumbai 400002, India +91 22815218 / 22815372 +91 22 2281 3200 info@thomasbaker.com, akshat@thomasbaker.com
4	M/s Thermo Fisher Scientific India Pvt. Ltd., Mumbai	Thermofisher	403-404, Delphi "B" Wing, Hiranandani Business Park, Powai, Mumbai - 400 076 Telephone:91-22-66803000 Fax:91-22-66803001/02
5	M/s Sigma Aldrich Chemicals Pvt. Ltd., Bangalore	Sigma	Plot # 12, Bommasandra-Jigani Link Road, Bengaluru - 560 100 Phone: +91-80-6621-9600 Fax: +91-80-6621 9450 Email: customerserviceindia@sial.com

Annexure – III

List of some suppliers of plasticware

Sl.No.	Name of the manufacturer	Brand	Details
1	M/s Tarson Products Pvt. Ltd., Kolkata	Tarson	Jasmine Towers, 2nd Floor, Suite NO.213, 31 Shakespeare Sarani, Kolkata, West Bengal 700017 Phone: 033 2289 2952
2	M/s Axiva Sichem Biotech, Delhi	Axiva	C-8 & 9, Ambey Garden, Libaspur, Samaypur Extension Industrial Area , Delhi -110042 (INDIA) Ph No- +91 8800493923 Email.: sales@axivasichem.com, Email.: anil@axivasichem.com
3	M/s Hi-Media Laboratories Pvt. Ltd., Mumbai	Hi-Media	23, Vadhani Industrial Estate, L.B.S. Marg, Mumbai - 400 086, India. Phone : 91-22-2500 1607, 6116 9797 Fax : 91-022-2500 5764, 2500 2468, 2500 2286 info@himedialabs.com
4	M/s Genaxy Scientific Pvt. Ltd., New Delhi	Genaxy	C-8 & 9, Ambey Garden, Libaspur, Samaypur Extension Industrial Area , Delhi -110042 (INDIA) Ph No- +91 8800493923 Email.: sales@axivasichem.com , Email.: anil@axivasichem.com
5	M/s Sachdev Scientific Equipments, Raipur	Kasablanka	Harchandrai Harimal, Dall Mill, Near Overbridge, Telghani Naka, Dist. Raipur (Chhattisgarh) 492001 Phone No. 0771 4265400, 0771 2239100 Email: info@kasablankacorp.com, sachdevscientific10@gmail.com, aditya@kasablankacorp.com

Annexure – IV

List of some suppliers of glassware

Sl.No.	Name of the manufacturer	Brand	Details
1	M/s Borosil Glass Works Ltd., Mumbai	Borosil	1101, Crescenzo,G-Block,Opposite MCA Club, Bandra Kurla Complex, Bandra (East), Mumbai - 400 051. IndiaT : +91 (022) 6740 6300 F : +91 (022) 6740 6514 Email : borosil@borosil.com
2	M/s Jyoti Scientific Industries, Gwalior	Jyoti	MAHADIK BADA, NIMBALKAR KI GOTH NO. 2,, Gwalior, Madhya Pradesh-474009 Phone : 0751-2322363
3	M/s Glassco Laboratory Glassware, Ambala Cantt., Haryana	Glassco	Manglai, P.O. Khudda Kalan (65.86 mi) Ambala Cantt, India 133001 Email: info@glasscolabs.com Phone: 0171 289 1633; 0171-6451250
4	M/s Riviera Glass Pvt. Ltd., Mumbai	Riviera	Subramanian Jayaraman(Coordination Manager) No. 2116, Oberoi Garden Estates, C Wing Chandivali Studio Road, Andheri East Powai, Mumbai - 400072 Maharashtra, India08048718895
5	M/s Sabar Scientific, Ahmedabad	Sabar	I-148/B, Phase II G.I.D.C. Estate, Vatva, Ahmedabad, Gujarat 382445 098250 23602

Zeitfracht Medien GmbH
Ferdinand-Jühlke-Straße 7
99095 Erfurt, Deutschland
produktsicherheit@kolibri360.de